Black Powder and Hand Steel

BLACK POWDER
AND HAND STEEL

Miners and Machines on the Old Western Frontier

by Otis E Young, Jr.
with the technical assistance of Robert Lenon

UNIVERSITY OF OKLAHOMA PRESS : NORMAN

Books by Otis E Young, Jr.

The First Military Escort on the Santa Fe Trail (Glendale, 1952)

The West of Philip St. George Cooke (Glendale, 1954)

How They Dug the Gold (Tucson, 1967)

Western Mining (Norman, 1970)

Black Powder and Hand Steel (Norman, 1975)

Library of Congress Cataloging in Publication Data

Young, Otis E 1925–
  Black powder and hand steel.

  Bibliography: p. 188
  Includes index.
  1. Mines and mineral resources—The West—History. 2. Miners—
The West. 3. The West—Social conditions. I. Lenon, Robert, joint
author. II. Title.
HD9506.U63A178   978'.02   75–4634
ISBN 0–8061–1269–7

  International Standard Book Number: 0–8061–1269–7

This book is for my dear wife,

RUTH KROST THOMAS YOUNG,

who has provided years of help
and moral support, sometimes under the most trying circumstances, and has
given me as well two tall sons, Otis III and Ben,
of whom I am enormously proud.

Preface

MINING men on the old western frontier were highly skilled artisans, proud of their admirable profession and much given to talking about it. When the desert-rat prospector returned to settlement after months of solitude, he had a great amount of conversational time to make up. The miners themselves were usually Cornish or Irish, two ethnic stocks which have seldom been regarded as taciturn. Unfortunately it is now too late to prospect for firsthand oral history of their deeds. Most of the old-timers have long ago gone to that Great Development in the Sky, excepting a tiny handful who are still living in secluded coves usually not far distant from the silent, skeletal headframes. Their story accordingly has to be pieced together as best it can from documents. The formal technical studies must be winnowed carefully for their human implications, for while the engineers' memoirs tend to be fascinating, they also tend to be edited so as to avoid scandal.

Many people and institutions have contributed heavily to this series of chapters whose common denominator is life and labor on the mineral frontier of the old West. Foremost among these is my longtime associate, Robert Lenon, P.E., of Patagonia, Arizona, a consulting mining engineer of vast experience and patience, as well as a fountain of wisdom and anecdote. Bob has read, expertly criticized, and added much of great value to this and all my other offerings to the field of American mining history, and I am deeply in his debt.

Professional mining men tend to be shy when first accosted, but when convinced of one's sincerity, they open up like a clam in the hot sun. A kinder and more helpful class of people I have never encountered. My sincere thanks go to Verne McCutchan, State Mine Inspector of Arizona and his able deputies; Walter Statler, Registered Assayer of Humboldt, Arizona; the engineering staff of the Magma Copper

Company, Superior, Arizona; Frank Ferguson of the Mollie Kathleen Mine, Cripple Creek, Colorado; Mr. and Mrs. Martin Duffy of the Florence Mine, Goldfield, Nevada; Mr. and Mrs. Ross Thomas of Dolores, Colorado; Leonard McCloud of Chula Vista, California; and Mrs. Martha Hickey of Mayer, Arizona.

Academicians to whom I am indebted include William Axford of the Charles Trumbell Hayden Library, Arizona State University, who obtained for my benefit some remarkably expensive documentary material; the administration of the Arizona State University for a sabbatical leave which permitted travel to some of the camps mentioned— mining history, like military history, goes much better after a personal view of the field; my esteemed and admired department chairman, Paul Hubbard, whose encouragement and interest have smoothed the path of research and writing; and the staff of the Bancroft Library for lending several critical microfilms. The fraternity of mining historians is small and cooperates freely, but I must single out for special mention Professor Clark C. Spence of the University of Illinois, Philip R. May of the University of Canterbury, Christchurch, New Zealand, and Watson Parker of the Oshkosh division of the University of Wisconsin.

On the editorial side I wish to express thanks and appreciation to Doyce B. Nunis, Jr., who supervises both the *Southern California Quarterly* and the *Brandbook* of the Los Angeles Corral of the Westerners; Don Bower and T.H. Watkins of the American West Publishing Company; and Harwood P. Hinton of *Arizona and the West.*

Otis E Young, Jr.

Tempe, Arizona

Contents

Illustrations

Color

Black and White

Map

Black Powder and Hand Steel

CHAPTER I

The Western Miner

FOLK songs to the contrary notwithstanding, the miner of the western frontier was in all probability not a forty-niner. Neither was he a prospector, although many desert rats of the old stock were drawn from the ranks of the argonauts of the California gold rush. Until about 1880 the miner was unlikely to be anything but a Cornishman or an Irishman, newly immigrated to the United States. When after 1880 the Celts began to drift away from the stopes, their place was taken by other new immigrants: Italians, Swedes, Finlanders, Austro-Hungarians, and Mexicans. All these ethnic stocks save the Irish had long and honorable homeland traditions of mining, so that today a modern copper camp in the West has a population which is a tin of mixed biscuits. It would be idle to suggest which of these groups was the most skilled. Each stoutly maintains that it is the Chosen. But viewed more or less objectively, all are very good. Underground work, like underseas war, quickly separates the skilled from the incompetent and by approximately the same method.

The Cornish came first, in the 1840's. They had placered and mined tin and copper in Cornwall from time out of mind, and the highest praise they could bestow on a mine captain (superintendent) was, "He do knaw tin." During the reign of the Tudors, Saxon technicians taught Cornishmen to hoist, smelt, and refine Cornwall's extensive copper deposits so effectively that for the next three centuries England held virtually a world monopoly of these two essential metals. Historians have made much of the English wool monopoly, which contributed so greatly to the kingdom's wealth during the Middle Ages but have had far less to say about the Stuart and Hanoverian prosperity, which was founded as much on Cornish mining as on wool production.[1] Perhaps this omission stems from the fact that expansion of grazing acreage drove men off the land through the enclosure movement, so causing poverty and social unrest, whereas mining provided work and reasonable contentment, hence eliminating the tumults and unrest which make for lively historical research.

Swansea, in southern Wales, just across the Bristol Channel from Cornwall, was the Pittsburgh of England long before the "black country" of the Midlands rose to obnoxious prominence. Swansea was just such a natural port as the Chicago-Gary complex, and it thrived for much the same reasons. The relatively calm waters of St. George Channel facilitated the economical transportation by sea of such high-bulk, low-unit-cost raw materials as Welsh coal, Cornish ore concentrates, China clay, and Irish foodstuffs, funneling them to Swansea, where highly skilled workmen took them in hand. For that matter, as late as 1870 it was considered more economical to ship Arizona ore concentrates by Cape Horn windjammers to Swansea than to freight fuel to Arizona by wagon in order to smelt the ore at the mine head.

Beginning about 1840, and repeating in 1865, Cornish mining prosperity slumped disastrously for a variety of technical and economic reasons. Most of the smaller lodes had struck water, and it is a mining truism that when the water runs in the dividends run out. The costs of deeper sinking and mechanical pumping in the larger lodes narrowed their profit margins to the vanishing point. The mine operators were very conservative and too often preferred to close down rather than experiment with technical innovations which might possibly have kept the setts active. Tin mining continued on a reduced scale, but in 1866 the copper industry crashed so disastrously that it was called the "copper collapse." The discovery of rich overseas copper deposits and a degree of mismanagement worsened the situation, throwing Cornish miners by the thousands out of work. In Victorian England the sole alternative to the unbearable social-religious humiliation of "going on the parish" was to emigrate. The Cornish miners swarmed to seek work in the United States, where the Alleghany coal measures, the Ozarkian lead deposits, and the Michigan copper range were beginning to be exploited on a massive scale and where skilled miners were in great demand.[2]

About the same time as the Cornish depression of the 1840's, another disaster struck across St. George Channel. Irishmen delved only for peat and potatoes, and when in 1846 the potatoes failed them, they could not eat peat. Their alternatives were emigration or starvation. The Paddies' transportation was in the steerage of the "coffin ships," where conditions nearly approximated those found in the slavers making the middle passage. Proportionally more Irish died on these voyages than had African slaves, for a slave showed no profit until he had been sold at Havana or Charleston, whereas an Irishman's fare was paid in advance; a bully skipper and bucko mates might therefore exert themselves to keep a slave alive, but not so an entirely expendable Irishman. Furthermore, social pressure in the South guaranteed the slave minimal standards of living, whereas an Irishman was obliged to root, hog, or die. In the natural order of things it soon came to pass that many an Irishman found himself on the end of a number four mucking scoop in an American mine drift, taking his orders from a Cornish shift boss. And like the shifter, he was probably putting away a bit of money to pay the passage of his wife and children to America. An "American letter" delivered in Cornwall was an invitation to a new life, but delivered in Ireland it was a license, perhaps, to live.

Both stocks are Celtic, marked as such by their black hair, blue eyes, and fair complexions. In their early days in America both were illiterate, the English establishment having found no reason to spend a sixpence on their education. Both tended to be garrulous, fond of the drink, uxorious, and inclined to the frequent reproduction of their kind. After this point, however, resemblance ceases. The Irishman was stoutly Roman Catholic of the Jansenist persuasion, whereas the Cornishman was devoted to shouting Methodism. The Irish were both skilled and radical in political affairs, whereas the Cornish were disinterested and conservative. Patrick would take any chance, however dicey, while John preferred to play his cards close to his chest. The Cornishman, or "Cousin Jack," as he came to be called, regarded mining as an admirable profession; to an Irishman it was only a way station to something better, like a seat in the United States Senate. Cornishmen thought wistfully of "'ome" and frequently returned, while the Irish passed the hat for revolutionary movements but had not the slightest intention of returning to the Ould Sod except as wealthy tourists. It would not be correct to say that they hated each other, for men who worked together in a mine drift could not afford that luxury. But it would be correct to say that they detested each

5

other's ways, fought like berserkers when in liquor, and kept their females insulated from the other's attentions.

The emigrant Irishman has been so often depicted—or caricatured—that little description of his appearance and ways is necessary even to a generation that has forgotten the meanings of "shebeen," "poteen," and "dudhean," three items once considered essential to the Hibernian life-style. The Cousins neither sought nor attracted much publicity, and it is only recently that their saga has begun to appear in print. Therefore, it is better to concentrate on the lesser-known Cornish, beginning with the fact their family names were unique. Children in mining camps quickly learned one version or another of the following couplet:

> By Tre-, Pol-, Pen-, and -o
> The Cornishmen you come to know.

In consequence wherefore, a Trewartha, Polweal, Penrose, or Trego was ethnically as identifiable as is any O'Riley, O'Grady, McCarthy, or Murphey.

In physique the Cornish were alleged to appear top-heavy, with shoulders and chests disproportionately large in comparison with their hips and legs, which seemed spindly. Perhaps this came from their profession, which placed great emphasis on arm and shoulder strength but little on walking. However, it seemed also to be true of their "Jennies" as well. An English observer, noting Cornish "bal-maidens" sitting on an ore platform and cobbing tinstone, remarked, "The use of hammers in dressing ores tends, perhaps, to the production of some fulness of breast, but the sedentary position necessarily gives little or no exercise to the lower limbs."[3] Hereditary influence should not be discounted entirely, for I have observed quite a few ladies of Cornish extraction who would not know a cobbing hammer if hit with one but who display fully both the garrulity and the physique traditionally associated with their ilk.

The dialect of the Cousins was as peculiar as their names and appearance. It was "shire" English, religiously omitting the verbal *h* where written grammars demand it and inserting it at the commencement of every word beginning with a convenient vowel. The favorite pronoun of emphasis was *she* whether the gender of the object suggested it or not, while *un* ("one") and *ee* ("thee") were nearly as common. When the Cousin was asked if the post appeared to be plumb, he squinted sagely and replied, "Damme, old son, 'ee's a bit more'n plumb. Putt they bleddy wedges to she, right 'ere." And two Cornishmen discussing a suicide were overheard to observe:

'Ee warn't waitin' fur me when I come by. 'Ee was 'anged op be the neck in the staable.
Did 'ee cut'n down?
Noa, I dedn' cut'n down. 'Ee warn't dead yet.[4]

The Cornish cuisine embraced a host of such dishes as kiddley broth, marinated pilchards, fermades, limpets, saffron boons (buns), figgy hobbin, and junket and cream, but the mainstay of the working miner was his pasty. This was a meat-and-vegetable pie enclosed in a football-shaped crust, "built" by his wife or the lady of his boardinghouse and issued him as he went out the kitchen door to the headworks. It fitted neatly into the lunch bucket and could be consumed without need for any table service but hands and teeth. When well prepared, it was as welcome as, and often referred to as, "a letter from 'ome," and it made a solid one-dish meal when washed down with a quart of hot tea. To this day restaurants in predominantly Cornish communities include a pasty of more moderate proportions on their menu selections, and its presence is a good indication that the town was probably once a mining camp.

The argonauts of 1849 may have included some wandering Cornishmen from Illinois or Pennsylvania, but they surged westward in great numbers following the discovery of the copper and iron mines of upper Michigan and the gold lodes of the Colorado Front Range. These demanded underground work, and the Cornishmen "knew better than anyone how to break rock, how to timber bad ground, and how to make the other fellow shovel it, tram it, and hoist it."[5] The other fellow in most of these cases probably hailed from Galway or Cork, since mucking, tramming, and hoisting by muscle-powered windlass are occupations which required less skill than strength. As the American mineral deposits were opened, Cornish setts continued to close, pushing still more skilled miners across the Atlantic. Indeed, their nickname is generally said to have arisen from the fact that whenever

The half-hour lunch break. The crewmen are indulging in Cornish pasties, seated on the reclining planks that serve as the miners' chaise longues. Timber wedges are their pillows. Reproduced by permission of Buck O'Donnell and Shaft and Development Machines, Inc.

a mine captain expressed desire to hire more men everyone in his crew offered to send for his Cousin Jack from Redruth, St. Just, or Truro.

Single men crowded in wilderness camps without family or other stabilizing influences are often prone to drink, and drinking, to rage. On such occasions the differences between Cousin and Paddy could flare into such a general brawl as one that occurred at Rockland, Michigan, in 1857:

> Richard Kestle was in a fight with two Irishmen, when a fellow Cousin Jack told him not to strike as there were only two of them! One of the Irishmen drew a long knife. Another came up with an axe and felled a Cornish fellow named Terrell, practically severing his back. Terrell was taken into Thomas's bar-room—a Cornish saloon—where he bled to death. At that the Cornish went wild: several score of them gathered and surrounded Ryan's grocery [a frontier euphemism for a saloon]. They smashed the windows and set fire to the building; Ryan leaped from the roof and disappeared into the wilderness— forever. The Cornish thereupon drove the Irish out of Rockland. At Portage Lake four hundred Irish swore revenge and set out for Rockland, where the outnumbered population took refuge on the steamer *Illinois*. But the Irish never turned up; something had distracted their attention: they fell to drinking on the way.[6]

With the passage of time and exposure in the melting pot, relations improved to the point where at Butte, Montana, the Cousin Jacks helped the Irish celebrate Saint Patrick's Day, and the Irish reciprocated on Saint George's Day. Even then the Irish genius for political stratagem was still resented. One old Cornishman was heard to complain, "Thee robbing H'irish, they not h'only 'ave two votes h'each on H'election Day, but the buggers vote seven years h'after they 'ave been dead h'and buried."[7]

However much the Irish manipulated matters "on the grass," the Cornish ran things to suit themselves in the stopes. Cornish inspiration may have lain behind the observation that wheelbarrows were dispensations of Providence, inasmuch as they taught Irishmen to walk upright. Thereafter, wheelbarrows were always the "Irish buggies" used for dumping ore down winzes (an opening between two levels in a mine) and chutes by those who had such matters in charge. Shifters tended to be brief in their instructions to newly landed Paddies. At Bisbee, Arizona, a shifter in the Copper Queen put one new Irishman to tramming and dumping down a winze and his friend in a lower level to pulling the same ore chute and tramming the filled cars to the hoist station. To the one the shifter said only, "Keep that hole full," and to the other, "Empty that chute."

The Cousins reserved the more skilled, but not necessarily the easier or less dangerous, work for their own kind. Aside from mucking, perhaps the most toilsome labor in any early mining operation before 1875 was drilling the holes for the drift rounds, and here the Cornish came into their own. "Double-jack" drilling was done by two-man teams, one man striking with an eight-pound sledge at the drill steel held, rotated, cleaned, and changed by the other member of the team. The stroke was a steady fifty beats a minute, and as a drill dulled, the man turning extended a finger as the signal to stop. The steel was pulled and changed, and the members of the team exchanged positions. The holder had to have great confidence in his mate, for a broken arm or a ruined hand was the price paid for an instant's inattention in the candle-lit gloom of the drift, where the target was only a bright spot the size of a quarter dollar. A mucker might well stay a mucker until he was incapacitated by the inevitable hernia and ruptured spinal disks; if he survived so long, he was probably crippled and worn out by his fortieth year. A driller on a face crew might rise to better things, but he had to rise fast before silicosis or "miner's consumption" carried him off. And if not silicosis, then the dangers of hang-fire drift rounds, caving ground, rotten ladder rungs, frayed rope, and fire were omnipresent. The Cornishmen accepted these hazards with the fatalism of soldiers and like soldiers would take hair-raising risks to rescue a trapped crew, knowing that as they did so they would be done by if their own time came.

Men who feel helpless before a malign and arbitrary fate tend to become highly superstitious. It is a measure of the Cousins' competence and optimism that their superstitions were few and for the most part based on shrewd observation. They talked of "Tommyknockers," the wee folk who inhabited mines and played mischievous pranks but who also gave timely warning of danger to those who heeded them.[8] Since heavy ground settling on a tunnel set will make the cap timber pop or knock at a measured interval, the inference is obvious. The Cousins disliked the assignment of numbers instead of names to mine workings;[9] perhaps a number is harder to understand than a name when shouted under adverse conditions. They objected to riding buckets to more than shallow depth, insisting instead on descending by ladders. "Cornishmen do not like to hang from a rope," they said. Hemp does not stand up well to mining conditions, and lowering men to depth by hemp or manila is tantamount to attempted murder.[10]

The Cousin Jacks would not touch a tool as long as a surface-based executive was present. This was partly professional pride but occasionally a necessary declaration of independence. One of the most disliked western mining superintendents was James G. Fair, an Irishman who achieved fame as one of the developers of the great Consolidated Virginia silver deposit at Virginia City, Nevada. Fair practiced a brand of personnel management calculated to set his crews' teeth on edge. He fired Cousins wholesale, claiming that "these damned Cornishmen know too much" for the good of his own stock manipulations. Another example was his habit of innocently asking the loan of a pipe from some miner and then discharging him at once if the bowl happened to be hot.[11] Fire was one of the worst dangers to any mine, and there is no justification whatsoever for smoking in a stope, but Fair's kind of safety engineering was bitterly resented.

Until about 1875[12] the table of organization of a mining development was almost purely Cornish. Boss of the ranch was the captain (pronounced "cap'm"), the chief executive selected for his mastery of all phases of mining, his "nose for ore," and his leadership. There was no question of his rising from the ranks; there was no other way to become a mine superintendent, although a pernicious rumor made the rounds that unlicked cubs graduated from the Columbia School of Mines (established in 1864) were actually being considered for such majestic posts. The captain was responsible solely to the board of directors, and, like a Royal Favourite, he rose along with profits or fell in consequence of "Irish dividends."[13] His personal staff varied with the size and complexity of the development but included timekeepers, surveyor-draftsmen, storekeepers, and perhaps a mine assayer. Like a ship's captain he enjoyed privacy, all available luxuries, and total self-direction but shared with a master mariner the responsibility of direct personal leadership into hazard when disaster struck.

The next grandee, corresponding to an executive officer, was the

The mine blacksmith. The blacksmith is reforging the star bit of a primitive machine drill steel. Reproduced by permission of Buck O'Donnell and Shaft and Development Machines, Inc.

timber foreman. He was in charge of the physical plant of the mine from headframe sheaves to sump. He oversaw the installation and maintenance of hoisting and pumping equipment; directed the sinking and timbering of shafts, stations, winzes, and adits; had the supervision of transportation equipment, men, and animals; and held the general mandate to present the mine as a going concern to the captain. Since a good timber foreman had to know a great deal about rock and its habits in order to do his job effectively and economically, he was frequently summoned to conferences dealing with proposed exploration and operations by a captain who valued the opinions of his subordinates.

More or less autonomous were the boss blacksmith and the operating engineers who ran the hoisting and pumping machinery. They were proverbially crusty and opinionated, ruling their subordinates without guile, with only the captain empowered to override their decisions. As long as the smithy kept the drill steel sharp and the machinery in repair, its wastage of coal and materials was overlooked. The pump man was invariably an ancient Cornishman who claimed to have known James Watt and Thomas Newcomen in their cradles, nursed the cranky pump pitwork, and was as dexterous as a monkey in clambering about its hundreds of feet of rod, guides, rising main, "aitches," and other esoterica which only he could understand. The hoist engineer not only ran the hoist engine but lorded it over the cager—the attendant who rode the cages, loading materials "on the grass" and the cars of ore and waste down on the levels—and the toplander—who worked in the mine yard to supply the face crews and to deal with cars hoisted to the shaft collar and platform.

Each of the two shifts had a shift boss, the "shifter," who directed mining operations on a day-to-day basis, while very large mines interposed level bosses between the shifters and the face crews. Such a boss moved about freely, visiting every stope, working chamber, maintenance station, and locus of activity at least once during his shift. He laid out the crews' work, diagnosed and prescribed for problems, approved unusual requests, and stirred up those employees whose vision failed to rise above tally and exchange (pay) day. His robe of office was his denim jumper (he wore it continually, whereas the miner usually laid his own jumper aside while at work). His side pockets sagged with ore samples, and he carried a small notebook in which he wrote, among other things, the names of those who would be given their time and "sent down the hill, talking to themselves,"[14] for conduct unbecoming. A less drastic means of persuasion was his prospect pick: as the boss was inspecting the lower haulage way or skip-loading level, he cocked an ear for the steady rumble of ore coming down the chutes and winzes as the muckers dumped it far above. Failing to hear this comforting noise, he pounded with his pick on whatever iron piping led up to the loafers' level. This produced the effect that the city apartment dweller merely hopes to achieve by pounding on his cold radiator to stir up the manager in the basement. But unlike the city man, the shift boss had means to enforce his will.

At the beginning of the shift the members of the face crew carried their tools and bundles of steel into the working chamber, adjusted their lights, inspected the face for missed holes (shots ignited by the preceding crew which had failed to explode), and fell to work immediately. The muckers pulled empty ore cars forward to the end of the track and began to fan the ore pile into them, shoveling with D-handled scoops off the iron turnsheet flooring the chamber, onto the smooth surface of which the broken ore had been deposited by the drift round previously fired. They had been admonished, "Fill the heel of the shovel; the toe will take care of itself." The turnsheet made this possible, in contrast with the harder work of shaft sinkers, who, blasting their way downward, could not use a turnsheet and so were required to "muck off the rough" with round-point engineers' shovels. The larger chunks of rock were lifted by hand into the cars, however, after being patted with a sledge a few times.

Meanwhile, the miners chipped the drainage ditch beside the track with drift picks, cleared the face sufficiently to begin their drilling from atop the muck pile, and perhaps stood and blocked a new tunnel set of timber, tying it into its predecessors with collar braces and installing lagging where necessary. Timbering was hard work but was never begrudged; there are tales of stupid or greedy mine managers who were heard to remark that men were cheaper than timbers. When the word got around, all that such brutes could hire were alcoholic drifters who would soon wreck the operation through incompetence or justifiable malice.[15]

A jumper, a loosely fitting denim jacket with a blanket lining, was worn while a miner was timbering, although its nominal purpose was to afford protection against the sudden changes of temperature and humidity encountered while moving about in a larger mine. Because the shifter was always on the move, he wore his jumper continually,

Muckers dumping ore down a transfer raise. The mucker in the fore-ground is keeping the hole full. The mucker in the background is patting oversized ore down to manageable size with a sledge. Repro-duced by permission of Buck O'Donnell and Shaft and Development Machines, Inc.

but the face crews usually worked in undershirts or even naked to the waist. Mine timbers, however, were covered with bark and splinters (and in a latter day liberally besmeared with creosote), yet while being wrestled into place they had to be hugged to the miner's breast as energetically as was his bride. While brides are seldom splintery,

15

The machine man and the chuck tender. The single hose indicates that this is an old-fashioned "dry" machine drill. It is mounted on a column, but the track ends are too near the face; the round will tear up. Reproduced by permission of Buck O'Donnell and Shaft and Development Machines, Inc.

an eight-by-eight timber was as prickly as a porcupine, and the jumper was then donned as a measure of protection for the chest. The posts, or side timbers, were set up, nursed to a true vertical, and were carefully wedged into place with rock waste or the wooden wedges which were supplied by the sackful from the timber yard on the grass. The cap timber's lower ends were shallowly dapped (flattened) with an ax and then set atop the posts. Newly opened ground does not take weight at once, and so the cap did not need to be a tight fit. When the ground settled, it would be pressed firmly into place.

When the shifter put in his appearance, the leading miner accompanied him to the face. Drifts follow the ore, and it was the shifter's business to see that the drilling for the next drift round was so arranged that the blast would bend the drift heading into the most economically advantageous course. This could be done by sight if the pay ore was clearly distinguishable from the country rock but was otherwise planned by plotting at the portal or shaft collar a running series of assays of the ore as it was sampled (hence the need for a mine assayer). With a few blows of his pick or, after 1910, spots of soot from the flame of his carbide light, the shifter sketched the drilling pattern to be followed, always edging the axis of the heading toward the more profitable parts of the ore horizons.

Drilling for the next blast began forthwith, the drillers mounting the muck pile and bracing themselves securely in position. Although shortcuts usually were possible, a formal round began with drilling a series of three holes to form a converging triangle at the center of the face. When shot out, this would produce a reliever cavity or cuthole. Around this, about midway between the side of the drift and the edge of the cuthole, would go the reliever holes. When these were shot, they would squeeze about half the face down radially into the reliever cavity. At the top and along the sides of the face would go the edgers, which would continue and enlarge the work to its full dimensions. Finally, a series of lifter holes would go in along the foot of the face, after total removal of the muck pile had at last provided room to swing a double-jack. Drilling was weary work, and it was a relief to sweating men when the senior miner pulled the last steel, cleaned the hole with his copper miner's spoon, and said, "Deep enough." This phrase became mining-camp slang for "enough, finished, that's sufficient, I am through," and was used with equal facility to start a brawl or to terminate it.

It was absolutely essential that all shots of the same kind fire as nearly simultaneously as possible and in the proper sequence as a group: collectively, they constituted one formal drift round. A properly fired round would pull a trifle more than six cubic yards of ore weighing thirty-two tons and would advance the heading about three feet. Two rounds gave an advance requiring the installation of another tunnel set, these usually being spaced about five feet from center to center. In hand-drilling days formal rounds were seldom employed, but when machine drilling came in, each shift could fire a full round as its last duty before riding the cage to grass.[16]

Unfortunately, the hurtling shower of sharp-edged muck from the first shots occasionally would shred the fuses of the lifter and lower edger holes, causing either a misfire or, worse, a hangfire. Electrical blasting was introduced as early as 1871 as a remedial measure.[17] Wires, to be sure, could be shredded and broken as easily as could Bickford fuse, but when an electrical shot failed, it failed all over. It never left a smoldering ember to creep along a frayed fuse covering for hours until it reached the solid powder core and flared into new life just as the drillers were going to work above it. To reduce some of the danger, the chief miner, or lead man lingered when he could within earshot of the round as it was firing, attempting to count the reports. If his count came out short, he left a message in the level station for the next shift, warning them to be careful.

The advent of machine drilling and dynamite altered but did not eliminate the hangfire problem. With dynamite it was now possible to run the fuse to a detonator cap in the next-to-headstick of the charge, so that at least three feet of fuse was contained in and protected by the hole itself from the bursts of muck. With some care the miner could now cut his fuses so that the lifter fuses would be burning snugly inside their holes before the cut hole went—a feat impossible to achieve with black powder. One spark escaping from the fuse covering would ignite black powder prematurely. Misses still occurred, of course, but the new danger was that a machine man might accidentally ram his steel into a concealed missed shot.[18] This was "drilling into a miss," and the expression became generally understood to mean "a serious

18

situation, created by negligent inadvertence." By way of illustration, one might say that when at Tombstone on October 26, 1881, the Mc-Laury-Clanton faction felt that the Earp-Holliday gang was indisposed to fight the McLaurys were drilling into a miss.

Like the open-range cowboy or the windjamming sailor, the western miner wore working clothes which economy and experience made virtually uniform. His Sunday wardrobe was probably indistinguishable from a banker's, consisting of a well-brushed black suit, derby hat, and gold watch chain with Odd Fellows, Knights of Pythias, Knights of Columbus insignia—but then the miner no more than the cowboy wore his working clothes on occasions of state. In the stopes, however, the emigré Cornishman was topped by a "hard hat," a felt headpiece of rounded crown and narrow brim, stiffened with resin or linseed oil until it approximated the modern construction helmet in its ability to deflect light blows. It had no formal provision for attaching a light, but a lump of clay held a candle at the front when needed; many an ambitious young miner put in his shift and then retired to bed to study geology by the light of the candle perched on his hard hat. Beneath it he occasionally wore a linen skullcap. Experiments were made with leather safety helmets like a modern fireman's or with sheet-iron flanged helmets, but the modified felt proved to be the universal favorite.

Next to the skin the miner wore woolen "union" undergarments conventionally long in the leg but short sleeved and fastened at the neck with two buttons so that the top could be opened wide and pushed down around the waist when particularly humid conditions were encountered. Over this went the jumper. Below the equator he wore trousers of baggy construction. These were "stagged," or roughly cut off a bit above the ankle to reduce the length that could accumulate mud—it is very easy when walking down a drift to put a foot into the ditch alongside the car tracks, baptizing the shoe and depositing a heavy ring of mud on an overly long trouser cuff. A few men tied the trouser leg just below the knee with a length of Bickford fuse (called "Yorks") in the manner of British agricultural laborers. Boots or brogans completed the ensemble.

Drifting or stoping was considered far more desirable work than shaft sinking, sometimes called "mucking into the bucket." About the

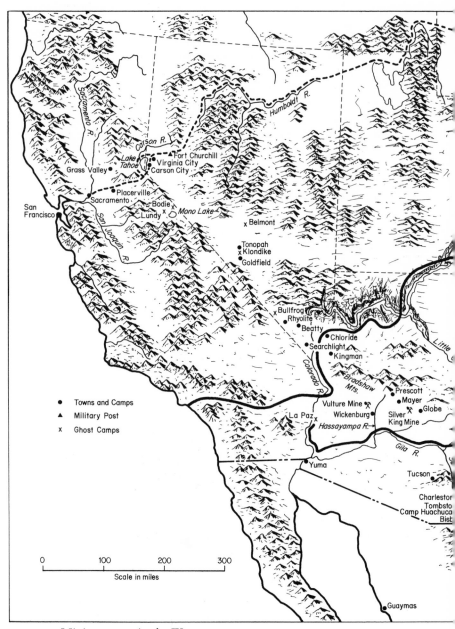

Mining camps in the West

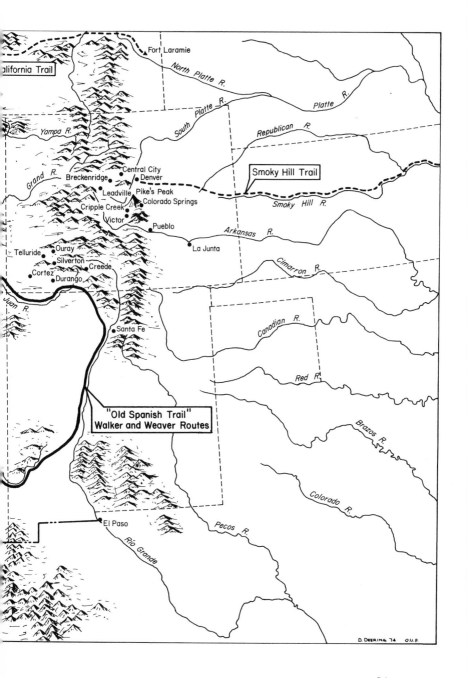

California Trail

Fort Laramie

North Platte R.

Platte R.

South Platte R.

Yampa R.

Grand R.

Republican R.

Central City
Breckenridge
Denver

Smoky Hill Trail

Leadville
Pike's Peak
Colorado Springs

Cripple Creek
Victor

Smoky Hill R.

Pueblo

Arkansas R.

La Junta

Telluride
Ouray
Silverton
Cortez
Creede
Durango

Cimarron R.

Juan R.

Santa Fe

Canadian R.

Red R.

"Old Spanish Trail"
Walker and Weaver Routes

Brazos R.

El Paso

Pecos R.

Colorado R.

Rio Grande

D. DEERING '74   O.U.P.

21

only redeeming feature of shaft sinking was that all the holes drilled were "down" holes, allowing the drillers to assume the most comfortable position. On the other hand, the difficulties were legion. The miners had to muck off the rough. Ventilation might be bad in the absence of natural air circulation, causing powder-smoke headaches from smelly muck piles. All muck had to be deposited by shovel or hand into the high-rimmed sinking bucket, and such a bucket was very difficult to pull aside to the shaft collar and dump when it reached surface. There was usually a steady deluge of ground water or condensation, requiring the wearing of uncomfortable oilskins and filling drill holes and dampening powder. Even though the shaft was lagged as much as possible, there was danger of loose rock, muck from the bucket, or even tools falling on someone's head. After splitting the fuses of a round, the miner had no way to avoid the consequences of the blast but to climb into the bucket and hope that the hoist engineer was paying attention to the "one bell" signal which in this case meant "hoist like hell!" Shaft sinkers were usually paid 50 cents more than the standard $3.50 a day, but mining syndicates more often preferred to let this work on contract to informal syndicates of miners rather than to undertake it with their own resources.

A peculiar aspect of the American social scene in the period of frantic industrial growth following the Civil War was the appearance of the itinerant worker who refused to put down roots however promising his current job but preferred the wandering life. They called themselves "boomers" or "hobos," and affected a caste system which clearly distinguished one of their own from the vagrant bum, who was usually an alcoholic mendicant fleeing from work as from the plague. Boomers prevailed especially in the West, where they were attracted to railroading, construction, lumbering, and mining. Such was their freemasonry that transportation seldom cost them anything; the average railroad "shack," or brakeman, was unlikely to evict from their side-door Pullman a group of his fellow hobos. Such aristocrats of the life as telegraphers "rode the plush," that is, they traveled on a pass or paid day-coach fare, but most of them preferred the less formal conditions found in a boxcar with a six-inch carpeting of straw. They dined on mulligan stew cooked over coal unwittingly furnished them by the railroad, and they professed mild disdain for the settled "home guard" who were tied to wives and children. It came to pass that eight men were thought necessary for a two-man job underground, since "Two were a-coming, two were a-going, two were a-rustling, and two a-working."

The migratory V's of Canada geese drifting south through the skies was a sign to many a mine payroll clerk that he would soon have extra work. The story is told of an old boomer who was laboring in a mine yard with a younger man and paused as the distant honking came to his ears. The old-timer laid down his shovel and remarked, "Them geese is going south where it's warm. And, Bob, I'm smarter than any dam' goose." Whereupon he rolled his sugan-wrapped bindle, accosted the timekeeper, and was on the next southbound freight dragging out of the camp.

Despite his flightiness, the tramp miner could be a good worker when he wished to be, and even formal textbooks spoke of him with a mixture of condescension and envy:

> The ten-day miner . . . worked with reasonable diligence at one job until he had accumulated a stake large enough to take him to another camp. Then he would either voluntarily "call the old hole deep enough" and quit; or he would soldier on the job until the foreman "gave him his time." Many such itinerant miners, competent when they wanted to be, made regular circuits: Butte, the Coeur d'Alene, Utah, Nevada, Arizona, and back to Butte.[19]

The home guard accepted the tramp miner's good-natured teasing, for the tramp was often an able consultant when some new and totally unexpected problem was encountered in the stopes. He would scratch his head, recall how a similar difficulty had been overcome elsewhere, and, with a workable solution presented, the problem was virtually solved.

Tramp miners began to fade from the western mining scene after the turn of the twentieth century. Unionization and technical change conspired to render his caste obsolete. Mine management found that Italians, Scandinavians, and Austro-Hungarians were (at least temporarily) cheaper and more docile workers. There was a plentiful supply of graduate mining engineers. The old-fashioned miner drifted away from the camps to find other work. Today, sixty years later, only a few are still to be found, living in secluded little spots throughout the West, and within another decade there will be none at all.[20]

CHAPTER II

"Five Bells!"

'5 Bells: Blasting or ready to shoot. This is a caution signal and if the engineer is prepared to accept it he must acknowledge by raising bucket or cage a few feet then lowering it again. After accepting this signal engineer must be prepared to hoist men away from blast as soon as the signal "1 Bell" is given and must accept no other signal in the meantime.' [Arizona State Code of Mine Signals]

ONE of the greatest of civilizing inventions is powder. It is essential to the winning of metals, without which mankind would quickly revert to a simianlike existence. It makes the strait wide and the crooked straight and in sufficient quantity requires even the eternal hills to bow down. In time to come its technological offspring may propel man to the stars. The uses for it are innumerable, not the least of which is the periodic suppression of great evil, to the benefit and betterment of humanity in general. That the ill-disposed may employ it on their own unlawful endeavors is a demonstration of the fine neutrality of the laws of nature, although it has been repeatedly observed that those who wantonly take up powder usually perish by powder, with hemp accounting for any leftovers.

There is no more pinpointing the origins of powder than there is in ascertaining who it was who first tamed fire. Scholarly research has heavily discounted the claims of Friars Roger Bacon of England and Berthold Schwartz of Germany. That the Chinese had something closely approaching it before A.D. 1000 is undeniable, but it is equally undeniable that they failed to grasp its potential. It is an open question whether their formula was transmitted to the Christian West or was there independently discovered. It could be conjectured that since both niter and sulphur were of engrossing interest to medieval alchemists and that they routinely employed charcoal in their furnaces, it is quite possible that an alchemist (or more probably his wife) in a fit of tidiness ordered the windrows of spillings on the floor swept up by an assistant, who then tossed the sweepings into the furnace. What followed would be of enormous interest to the master, whose first thought could be that an instant depilatory had been discovered.

Europeans puttered about with powder for some decades, exploring its military possibilities in obedience to the maxim that soldiers who are grasping the repugnant end of the tactical stick will also grasp for any technical innovation which shows promise of becoming an effective equalizer. Accordingly, the first serious appearance of powder was on the battlefields of the opening rounds of the Hundred Years' War. The vastly outnumbered English deployed *pots au feu* or crude, mortarlike cannon, whose fire mission seems at best to have been to stampede the horses of the overwhelmingly numerous French feudal array. The firepots were quietly retired when the English discovered to their amazement that their disregarded archers were capable of ruining any number of armored knights. These guns proved equally uneconomical in battering French strongholds, which fell much more readily through a combination of psychological warfare and political intrigue. Within two generations the English had the French on the run, whereupon the French themselves picked up gunnery and by 1400 were developing it as an effective weapon of war.[1] Joan of Arc's minor tactics neutralized the British arrow storm, and Jean and Gaspard Bureau's siege trains ripped open the English-Burgundian strongpoints. Within a miraculously brief period the English were expelled from France, with the French themselves becoming committed to artillery as the weapon of military decision.

The gunpowder formula and the principle of the gun spread through the Western world with speed and were next seized by the Ottoman Turks, who were having difficulties dealing with the Christian concept of defense in depth by means of fortified strongpoints about which lightly armed Asiatic hordes swarmed vainly. The gateway to southeastern Europe was Constantinople the well fortified, 'Mikkelgard' to the Vikings. For strategic, economic, and dynastic reasons the Osmanli sultans had to lay hands on this prize. It was so tough a nut to crack, however, that it had been successfully stormed but once before, and then success came only with the help of surprise and a degree of treachery. The renegade Hungarian gun founder Urban now cast Sultan Mohammed II some enormous bronze siege pieces. After a brilliant amphibious campaign brought the Turks within effective range of the walls, Urban's huge guns battered breaches through which the Turkish storming columns surged to take Constantinople in the ill-omened year 1453.

As committed now to guns as were the French, the Turks made good their hold on the eastern Mediterranean and the Balkans and in 1529 marched northwest against Vienna, the keystone to the defense of Germanic Europe. If Vienna fell, everything west of the Rhine would go. Suleiman I, "the Magnificent," succumbed to hubris, however, and refused to allow a phenomenally wet summer to interfere with his timetable of conquest. The great siege pieces could not move over the foundered roads or up the shallow rivers, whereupon Suleiman employed Wallachian and Moldavian miners to sap the walls of Vienna, intending to breach them with enormous gunpowder mines. His miners sapped the alluvial ground expertly enough, but the defenders of Vienna discovered what was going on and countermined as though their lives depended on it; in this case, and in view of Suleiman's way with prisoners, this was literally true. So defective was the Turks' ignition method that Suleiman could not prevent the Viennese from entering and robbing his powder chambers after they were charged. This not only neutralized his breaching technique but resupplied Vienna with cannon powder for its wall pieces. Three unsupported Turkish storms were turned back from the comparatively undamaged walls with terrible losses. Turkish morale cracked, and Suleiman withdrew, initiating the long slide of his realm to decrepitude. It might almost be said that the Ottoman Empire began to perish for want of a reliable fuse.[2]

At this point the history of military demolitions and propellants parts company from engineering explosives, not to converge again until the static siege warfare of World War I revived military mining on a scale which Suleiman I would have envied. The Renaissance soldiers concentrated on their guns and petards, striving always to increase the mobility and shock effect which are essential to any effective weapon, while civil engineers began to look at powder from the viewpoint of their own professional requirements. Though not then expressed as such, powder represented man's first high-order "power package" capable of operating in reasonable independence of its environment. It had the additional attractions of being cheap, being flexible in employment, usable by the average worker, having few bad side effects, and producing results fairly proportional to the quantity employed—none of which, by the way, are true of nuclear energy. Its most obvious and ideal application lay in shattering rock. It can be generalized that scarcely any engineering project of major proportions can be accomplished without first founding it on or within solid rock; yet preparing such a foundation has always constituted one of the chief items of over-all project cost. Any technique which could economize in money or time in this respect would have a major impact on human development.

Man had founded on or burrowed within hard rock for thousands of years before the invention of powder but had done so only by "cold mining" with human effort alone or by "firesetting," which involves breaking out rock by means of thermal stress induced by kindling a hot, prolonged wood fire against the face of the work. Occasionally, when the ground was exceptionally favorable, holes could be drilled in the face of the heading and stuffed either with quicklime or with tightly fitting wooden wedges. When wetted under proper conditions, these materials would expand with sufficient force to break out the rock. Even with slave labor, all these methods were exorbitantly expensive and lay within the capabilities of only very wealthy and highly organized states. Until the seventeenth century major civil engineering was carried on as a function of the state, supported by all its resources, with the state enjoying a monopoly of the benefits, whether military, religious, or economic. A brief scanning of the classical writers shows, for instance, that precious-metals mining was a state monopoly, save for those deposits (placers and the like) which lay in ground so friable that the work consisted more of excavation than of mining proper.

Nonferrous mining had been conducted throughout Europe since earliest times but had latterly been hampered by the disorganizations inherent in feudalism and by inadequate technology. One such mineral strike had been made in the Harz Mountains of Saxony about 1170, when deposits of silver lead were found and exploited at the grass roots.

These Saxon mines, like all others of their time, were relatively unproductive until about 1450, when the Renaissance knowledge explosion began to produce in that region a revolution in mining organization and technique. Within the century 1450 to 1550 modern mining had its real inception. The state did not retire completely from the scene (it never does), but princes found it more profitable to lease or grant mineral rights to private entrepreneurs, who then went after the ore with small, efficient organizations, well-thought-out methods, and moderate capitalization.[3] Even in newly conquered Mexico and Peru the Spaniards, after briefly experimenting with politically underwritten slave mining in the classical manner, quietly went over to free mining based on Saxon methods insofar as they proved applicable to radically different terrain and conditions.[4]

By around 1550 the arts of mining, milling, and metal refining had been so improved over medieval practice that the three great textbooks of the art not only could be written but would have far more relevance to modern methods than they had to techniques only a century or so older. These texts were Lazarus Ercker's *Treatise on Ores and Assaying*; Vannocchio Biringuccio's *Pirotechnia* on metallurgy; and the pre-eminent *De re metallica*, by Georg Bauer. Bauer's great work, published in 1555 shortly after his death, remains to this day an object of admiration among mining men, who esteem it as the fundamental canon of their art.

Still another century elapsed before powder was adapted from military to civil employment, the first use being in a grenadelike explosive-incendiary bomb tossed into headings of doubtful safety to bring down loose rock by concussion before it fell of itself on someone's neck. Additionally it would ignite or help expel any dangerous gasses which might be pooled in the workings. In 1627 one Kaspar Weindl took matters the logical step further by setting off the first intentional blast to pull rock. The presumption is that he poured his serpentine powder into handy cracks or crevices, tamped them in some manner, and fired with military slowmatch. The advantages of this process were so obvious that the practice had spread to Cornwall by 1689 and to the Swedish copper mines by about 1724. With powder one or two skilled free miners could break out far more ore than could any number of slaves or mine churls (serfs) inefficiently crowded into a narrow heading. This accelerated the trend to private enterprise, since even a lackwit can perceive that it will not do to acquaint unfree and resentful labor with the political possibilities of powder. Now and then these incommensurates have nevertheless been attempted.[5] They have invariably had the same outcome, lending support to Georg W. F. Hegel's pessimistic conclusion that all we learn from history is that we learn nothing from history.

In 1683, Hemming Hutman, of Saxony, invented the rock drill so as to place his blast holes to greater advantage. As first practiced, a hole of some fingers' width (about 2½ inches in diameter) was driven into the rock a sufficient depth—"sufficient" being a most variable and critical matter. The tools were a hammer and a drill forged of the best available wrought iron, with an inset bit of steel as well tempered as possible—by that time swordsmiths were well experienced at tempering and had no great difficulty in hitting upon the right steel heats and water temperatures. When judged deep enough, the hole would be cleaned of the cuttings and if need be "clayed," or lined with clay with a special tool, to exclude water and fill fissures. The hole was packed with conscientiously prepared powder, or, if seriously wet conditions were encountered, the powder was enclosed in a cloth bag impregnated with pitch. So far so good, but at this point miners would be balked for another two hundred years.

The question was ignition of the powder, a process which ideally should be reasonably safe, certain, and predictable. Military experience was of no assistance. The musketeer's slow match was only a loosely spun cord of tow impregnated with saltpeter, whose sole function was to maintain a coal of fire. Its burning time was slow and wholly erratic. The military engineer's demolition quick match was a loose hollow train of powder enclosed in a paper or parchment tube, burning like a flash. Neither would do for mining where dampness, abrasion, compression, and kinking could produce an annoying misfire or the more dreaded hangfire (an unpredictably delayed explosion). The miners of the Harz and Erzegebirge mountains evolved their own ignition system, called "plug shooting," based on a grooved wooden cone which carried the ignition compound in the groove. When hammered solidly into the collar of the hole, the plug also acted as stemming (a tamper) to prevent the shot from blowing away out the hole to little or no effect. These plugs functioned fairly well, but the groove was too short for safe delay and was incapable of permitting any variation in time of burning.

This meant that blasting had to be confined to putting down and shooting out one hole at a time. This method was faster than cold mining and cheaper than fire setting but far inferior to firing rounds— igniting a series of adjacent blasts in quick, sequential order. Shooting one hole with a plug might be done in reasonable safety, but attempting to shoot rounds in this manner merely courted disaster. Unable to find a way out of the problem, miners then made the mistake of drilling their holes too deep, thinking that the increased powder capacity thereof would offset the inefficiencies and inadequacies of a single blast. This theory proved illusory; too deep a hole would blow out a narrow cone rather than spall off a wide slab of rock, producing a "gunpocked" face on which it was extremely difficult to continue work.

Nonetheless, the ingenious Saxon miners improved what they had to the best of their abilities. In 1685, Bergmeister (Superintendent) Carl Zumbe, of the Clausthal mines, invented clay stemming, whereby clay, sand, or drill cuttings were used to stem the hole. Such material was superior to wood since it would expand as the blast was fired, plugging the collar more effectively. In the following year a Clausthal foreman, one Singer, introduced the priming tube of copper, wood, or reed filled with powder, bound about with string, and inserted in the charge. The clay stemming was then pushed and packed in about it to even greater advantage. Hans Luft, a bookbinder of the same district, in 1687 devised the paper powder cartridge, which ended the necessity of dealing with loose powder and made the loading of "up" holes easier. Finally, the drill hole was reduced to a more practical 1½-inch diameter and its length shortened to about 20 inches to prevent cone spalling.[6]
At this point the miners sensibly rested content, pending a major breakthrough in technique, although optimists continued to blow themselves up at regular intervals while attempting to find a safe way to shoot rounds.

Saxon mining methods were disseminated throughout most of the western European world through the works of Georg Bauer and by the emigration of Harz miners and millmen. Some of this hiving went in the direction of the British Isles, where the Saxons helped transform the placering and grass-roots open-cast winning of tin into the relatively deep mining characteristic thereafter of the Cornish and Devon stannaries. The geological structure of the Cornish tin lodes was unique: on a large-scale map the major deposits appear as a series of Greek crosses, suggesting that there were two distinctly different periods of mineral intrusion, one cutting in one direction and the other subsequently cutting at right angles thereto. The individual lodes are thin and narrow, frequently consisting of a series of parallel ore shoots, dipping almost vertically into the earth, like the pages of an upended book. These regularities made prospecting and subsurface exploration fairly easy. The country rock is an ancient granite, hard but without vices save for the presence of clay. This terrain was made for deep mining with powder, and the Cornishmen fastened upon the Saxons' devices with enthusiasm.

As the tin wheals sank deeper, the inevitable problems encountered spurred all manner of inventiveness. The British iron industry was levied upon to provide better and cheaper drill steel. Double-jacking was perfected. The Cornish began a more rational approach to the problem of ore transportation. Then, in 1831, William Bickford, of Tuckingmill, invented the first really effective fuse. He apparently drew inspiration from watching rope being spun, for his fuse was manufactured by twisting six strands of jute about a core of powder, wrapping the whole with strong twine, and then wrapping it again with one or two layers of tape made waterproof with tar or gutta-percha.[7] Bickford fuse burned almost exactly one foot in thirty seconds, was proof against most of the normal vicissitudes of mining, and when it failed tended to "fail-safe" more often than it contributed to hangfires. When produced by machinery, it was quite cheap. William Bickford was one of the chief contributors to the industrial age, for his critical invention made possible the firing of rounds that broke out rock or ore by the ton rather than, as before, by the hundredweight.

Miners quickly adopted Bickford fuse and learned to compensate for its few bad habits. Hangfires, caused by smoldering, muck-shredded fuse wrappings, simply had to be waited out, but this was best accommodated to by firing the round at the end of the shift. In the hour that intervened between firing and the arrival of the next shift, most hangfires would either shoot or give up the gun. Lighting the fuse itself constituted something of a problem. A fuse cut squarely across the end was rather difficult to ignite and had to be held in a flame some seconds before it took hold with its characteristic spit of sparks. By way of compensation, the miner split open the free end of the fuse about an inch, exposing more of the powder core to the flame; to "split a fuse" was such a standard preliminary to firing that it became synonymous with the ignition itself. The initial spit was also likely to blow out the candle flame—an awkward result in a day before the invention of sulphur matches and at a time when only one miner and one candle were present at the face. To offset this problem, the miner collected the "snuffs" (discarded candle ends) in the drift, lit them all, and used them one by one. Much later, mining companies would furnish a firework resembling an obese sparkler, which was proof against blow-out. Before use the miner carried it by bending the bare wire end into a hook and hanging it in a convenient jumper buttonhole.

The favorite ignition device, however, was the old reliable "spitter," a length of Bickford fuse cut shorter than the fuse rattails hanging down from the hole collars. It was notched down to the powder at intervals, one notch for each hole to be fired, plus a few in reserve. The spitter furnished a nice jet of flame for each rattail as the fire came to the notch. Likewise, the notches served as a convenient tally to ensure that no hole had been overlooked in the pressure of firing. Best yet, it had a built-in safety factor, for as it burned to the end, the fingers holding it would grow uncomfortably warm, and it would be automatically thrown down. This ended at once any temptation to hang around in the drift and forced the blaster to depart well before the holes began to go off.

With double-jack and steel, black powder and Bickford fuse, wire rope and steam engines, western mining entered into its salad days about 1860. This pattern held for ten years, modified only by the development of DuPont's "soda" blasting powder, in which the cheaper but no less effective sodium nitrate was substituted for the traditional potassium nitrate in the formula. It had not been practical before, since sodium nitrate usually absorbs so much moisture from the air that powder made from it ordinarily becomes too damp to go off. The DuPont Company ingeniously found a way around this difficulty and sold its product under various brand names throughout the West. The powder keg in which bulk shipment was made became the miner's table and chair, and the company brand on it was as accepted a part of his decor as was "Pillsbury's Best" across the seats of the lingerie of nesters' wives or "Arbuckle's Coffee" on the burlap bags in which cattlemen carried their possessions.

The work of loading holes of the drift round began as soon as the last lifter hole was cleaned. The apprentice tool nipper brought the box of powder cartridges and a reel of fuse to the leading miners. As they addressed themselves to the work, the rest of the crew retired to do such necessary but distant work as timbering, clearing away tools, and drainway chipping. The boss miner stood up to the face, ramming in the cartridges with a long wooden rod as his partner handed them up from the box and slit or twisted them open. Their coverings were thus ruptured so that the ramming could expand their contents to fit the hole snugly. With the last cartridge (later, with dynamite, the next-to-head cartridge) the fuse was inserted and firmly tied in place to prevent its pulling away. The hole was carefully stemmed with clay or drill cuttings flush to the collar. The miner then moved on to the next, leaving the fuse rattail hanging down the face. Loading a full round was a time-consuming task requiring the better part of an hour and a full measure of conscientious devotion to duty. Experience helped as well, since each charge needed to be varied slightly according to the type of rock and its condition to produce maximum effect without wasting powder or underloading.

With all in readiness the crew walked to the level station adjoining the hoist compartment. Someone summoned the cage by pulling smartly on the annunciator wire that rang the level code on the bell in the hoist room far above. When the cage appeared, five bells[8] were struck to warn the hoist operator that the crew was ready to shoot. When the hoist engineer jiggled the cage to indicate "understood," the crew so informed the blaster. Alone in the working chamber, he lit the spitter from a snuff, cried the traditional, "Firin' the hole!" and began to move from rattail to rattail, lighting them in order of their desired firing: cut hole, relievers, edgers, and lifters. As soon as all were smoking and sparking satisfactorily, he rejoined the crew. They all stepped into the cage, rang one bell, and were carried up and away from the imminent blast, as soon as the lead man had heard and counted the blasting holes.

Now and then there might be slipups. It was difficult in a shaft mine for a casual wanderer to gain access to an area about to be shot, but in an adit mine, where people went about on foot, they could find themselves in a blasting area. One miner found out about this the hard way:

One afternoon, not long after lunch, I was measuring the advances made by each contractor and . . . had passed the cross-cuts on my way to the drift face when I smelled smoke from lighted fuse. I should have known something was wrong when I did not see lights or hear men working. Like everyone else in the mine I had grown careless. . . . I crawled between the wall and an empty ore car, and stopped to call to anyone who might be nearby. There was no answer. At that moment a hole in one of the cross-cuts blasted and my candle went out. In a split second I had turned the ore car on its side and thrown myself into it. Then all hell broke loose. All three faces were blasting at the same time, the round in the drift and the rounds in the two cross-cuts. As usual the holes had been overloaded and rock from the drift face was catapulted against the bottom of the overturned car. Rock was thrown across the drift in both directions from the cross-cuts.

When I was sure that the holes had blasted, I got out of the car and, without light, crawled on my hands and knees to the main drift five hundred feet away, through powder smoke that was strangling and that almost made me tear my lungs out from coughing.[9]

This incident occurred when the miners were shooting with dynamite, but it could as easily have happened with black powder; Mark Twain remarked that it was poor consolation to a corpse, however, to know that the powder which blew him up was of inferior quality. He also touched on one form of carelessness with black powder that was the result of criminal negligence, by comparing Micronesians whose faces and necks were tattooed blue to "the customary mendicant from Washoe who has been blown up in a mine."[10] There is no doubt that the unfortunate miner witlessly tried to tamp his powder with an iron bar or even the butt of a drill steel. The iron had struck a spark from the rock, and the powder had flashed out the hole, tattooing his face and hands with unburned grains. For this very reason careful blasters used tools made only of wood and copper and were conscientious about keeping their lights far back from the face during loading.

The partnership of black powder and hand steel lasted in the West for perhaps fifteen years. Though the tonnages of ore they moved were enormous in comparison to that broken out by older methods, the bigger mining corporations writhed at the expense of payrolls, which included hundreds of double-jack teams. It was by now evident that drilling could and should be done by machines—the basic movements were few, simple, and repetitive, a formula which invariably suggests mechanization. Both in Europe and in America inventive minds turned to the idea of adapting the age's prime mover, the steam engine, to this work. As early as 1855 in Alpine railroad tunneling and in 1862 at the Hoosac Tunnel in Massachusetts machines originally designed for steam but converted to compressed air[11] were piloting the way. The European designs were cumbersome and fragile, but in the United States machine drills were stripped to the bare essentials, with few but rugged components, drawing-board efficiency being sacrificed to gain long life under actual working conditions.

The machine drill adopted in the West was an elongated and relatively slender double-action reciprocating air engine sold variously as the Burleigh, Rand, Ingersoll, Leyner, or Climax; all of them were similar in fundamental design. The valve gear was superimposed on the cylinder in the manner of a railroad locomotive. As air was admitted through the valve to the rear of the cylinder, it drove the piston, connecting rod, and drill chuck forward in the percussion stroke. At the top of the percussion stroke a knob on the piston's rear extension tripped the valve the other way, admitting a somewhat smaller shot of air in front of the piston to return it to the bottom of the cylinder. At that point the knob flipped the valve the other way, initiating the start of a second cycle. The second function, rotation of the steel, was accomplished by a spirally splined "rifle bar" and ratchet at the rear, the bar engaging the piston extension and turning the steel counterclockwise an eighth of a turn at the end of the cycle.[12] The third function, advancement of the steel as the hole deepened, was performed by the long screw jack on which the drill was mounted, the machine man turning the screw by feel and sound to keep the steel well up in the hole.

The two-man drilling team was preserved, but assumed entirely different functions. Two men were necessary to set the heavy, clumsy machine in position on the massive column or bar which supported it. An adjustable universal joint permitted the machine to be directed at almost any point or angle except for the lifter holes, which were both too low and angled wrong to give a bar-mounted machine proper clearance. Accordingly a "Finn board," or heavy plank trough, was used to secure the machine in this position, the team bracing the Finn board with their backs and feet. The machine man himself operated the drill, while his chuck tender stood between the machine and the working face. When a steel dulled, the machine man stopped the air and retracted it to allow the chuck tender to unbolt it from the heavy chuck and slip in another and longer steel. At this point he usually dashed some water into the hole to help reduce the dust created by this dry drilling. Matters proceeded in this fashion until the machine man said, "Deep enough," and the universal joint was loosened to slide the machine opposite to the point where the next hole was to be collared.

The first major use of machine drilling in the West was in the driving of the Sutro Tunnel at Washoe, Nevada, in 1874, although prototype Burleigh machines had been tried out in the Comstock's Savage Mine two years earlier.[13] The air supply was carried in common iron piping, delivered by simple one- or two-cylinder reciprocating compressors driven by steam. In Europe air compression was attempted by designs of vast ingenuity, involving collecting the air entrained and compressed in penstocks leading down from Alpine streams. These hydraulic compressors failed because their creators lacked materials that would stand up to the water hammer, which soon pounded the outlet valves to pieces.[14] American compressors wasted power but delivered the air reliably—unless lubricated with oil of the wrong quality. Should enough such oil become vaporized, compression heat would ignite it and blow the compressor sky-high—often enough taking a good length of the air main with it.[15] Humorists have in this connection suggested that the diesel engine was here struggling to invent itself but that no attention was paid to its birth cries except that the operating engineer was fired and a new compressor was ordered.

Birth cries of even louder amplitude were simultaneously being produced by the move toward high explosives. Chemists had found that they could transform simple organic molecules into interesting new compounds by treating them with nitric acid in various ways. In 1846 the Italian experimenter Ascanio Sobrero thus nitrated glycerin and then heated it to see what would happen. He found out very rapidly. Some practical work by chemists of great courage established that this nitroglycerin was up to eighteen times more powerful than an equivalent amount of blasting powder and that, because of its tendency to detonate rather than to burn, its shattering effect on rock was even more pronounced. Presently "blasting oil" was being commercially manufactured and shipped by the Nobel brothers, of Sweden. San Francisco, the entrepôt for the mining West, was the site selected by fate for the demonstration on April 16, 1866, that nitro is remarkably sensitive, particularly when manufactured and moved without special precautions. An involuntary observer told his story.

In 1866, I sailed from New York for San Francisco by way of Panama; when we reached Aspinwall we crossed the Isthmus to take the Pacific steamer at Panama [City]. . . . I was standing near the gangway when the baggage and express matter came on board, and I . . . assisted in taking a few of the boxes over the rail. . . .

[On April 16, shortly after 1 P.M.] I was walking up Montgomery Street . . . toward the Occidental Hotel. I had been there but a very few minutes before I heard a loud report, which jarred the whole building, and set people flying through all the corridors to ascertain what was the matter. I went out, and walked up the street the way I had come. The office of Wells, Fargo, & Co.'s Express was, if I remember correctly, two blocks away from the hotel. It turned out that the explosion which had jarred all that part of the city, was in the office of the Express Company. . . . Among the boxes which had been on the steamer with me from New York to San Francisco, had been passed over the rail of the steamer at Panama, and which I had assisted in handling, there were two cases of nitro-glycerine.[16]

The *Mining and Scientific Press* conducted a technical autopsy of this accident, which had occurred virtually on its doorstep. The explosive had been shipped in iron flasks bedded in sawdust within the wooden cases. The cases had not been labeled or marked with the nature of their contents, and at least one flask had leaked some of its nitroglycerin into the sawdust packing. After the cases had been stored in the Wells Fargo yard behind its office, someone noticed a leak puddle, attempted to pry or knock open the case, and succeeded in exploding the one, which promptly detonated its companion. Eight people were killed outright, and three died soon after of their injuries. To clinch the point, news came a few days afterward that the steamer *European*, docked at Aspinwall on April 3 with ninety cases of nitroglycerin loaded at Liverpool, had experienced the same mishap. Fifty people died of the "infamous oil."[17]

This evil habit of unseasonable detonation was overcome in the following year by Alfred B. Nobel, to his great glory and personal profit. He found that soaking some absorbent material, such as fuller's earth, with carefully purified nitroglycerin rendered the product resistant to jars, jolts, and shocks, and yet when properly detonated its power was but very little diminished. His new dynamite could be manufactured in any desired strength, 30, 40, and 60 per cent mixtures becoming the favorites. The moist compound was packaged in waxed-paper rolls or sticks which could be slit and loaded like blasting powder and were a great deal safer and easier to handle than liquid nitro. If subjected to open fires, the new "giant powder" would burn but would not detonate; for that matter it was so inert that special measures had to be taken to ensure that it would explode when desired. The solution was to devise fulminate-of-mercury detonators or blasting caps of short lengths of copper tubing closed at one end, into the other end of which Bickford fuse could be inserted and firmly crimped. The fuse spit would ignite the fulminate compound, and the shock of the exploding fulminate was in turn adequate to set off the dynamite. Blasting routines were little changed except in such details as hole depth and spacing, adjustment of charge, and priming of the next-to-head stick rather than the tail cartridge in the loaded hole. The DuPont "powder combination" was licensed to manufacture dynamite in the United States, and it sold in the West under various brand names, the most prominent of which was Hercules.

There is no rose without a thorn, no technological advance without its inherent drawbacks and problems. For all its virtues the new giant powder was no exception to the rule. Its fumes, for instance, were even more noxious than were those of black powder. It froze into ineffectiveness in cold weather. It was of little use in coal mining, being so powerful that it powdered the coal into slack. The detonators added a new and permanent element of hazard to mining-camp life, for they tended to get scattered about promiscuously, while their sensitivity and power have to this day continued to destroy the noses, fingers, and eyes of unwary souvenir collectors.[18] If allowed to stand in idleness for any length of time in warm storage, the liquid nitroglycerin began to flow down and seep out of the paper wrappings, collecting in puddles, crystallizing, and regaining its vicious sensitivity. Miners hated dealing with old powder. When they found it in abandoned workings, all they could do was drench it in water to dissolve and dilute the crystals, carry the cases away more tenderly than any baby, and burn it in the open air where it could do as little harm as possible.[19]

On the other hand, the power and flexibility in use of dynamite endeared it to the western miner, who learned to use it for tricks verging on the fantastic. One of these tricks involved making a drift muck pile flip itself backward on the turnsheet like a Japanese acrobat. This feat not only was admirable in itself but was of considerable value in economizing on a crew's time in mines, where the management was sufficiently alert to realize that two sticks of 40 per cent powder cost less than inefficient labor by the men. It probably originated in the observation that the drift round frequently did unspeakable things to the iron turnsheet, even though its forward edge had been pulled back a few feet from the face about to be blasted and had been anchored down with a few hundred pounds of muck. Despite these precautions the lifter blast often got under the sheet and rolled it up like a tortilla. Not only was the sheet ruined but the crew then had to muck "off the rough" ahead of it, in the undesirable manner of shaft sinkers.

Under the new dispensation a large piece of half-inch boiler plate was laid down in the drift. Its forward edge was tight against the face, while its back edge was beneath the turnsheet, leaving three or four feet of the plate exposed. Beneath the leading edge of the plate was placed a stick or two of 40 per cent powder with longer-than-lifter fuses. When the round had fired, it left the muck pile in temporary repose on the plate. At this moment the charges beneath it would go and would flip the heavy iron back, tossing the bulk of the muck pile to the rear. This left the face comparatively clear, ready for drilling to start within a few minutes after the arrival of the next shift. Since the muckers no longer had to monopolize the face for an hour while clearing it to permit work to begin, the crew had additional time to work together on arduous jobs such as setting up the drill columns and rerailing loaded cars.

The final frontier-period improvement to blasting was electrical ignition, worked out by the larger mines to minimize the hangfire problem and to secure absolutely simultaneous firing of a given set of holes. It involved little more than replacing the Bickford fuse spark with a bit of platinum resistance wire in the detonator. Electricity from a wet battery was then supplied through insulated wires, whereupon the resistance bridge heated to incandescence to fire the detonator and the charge. To economize on wire gauge and electrical supply requirements, these detonators were made sensitive to very small amperages, so much so that passing electrical storms and related phenomena occasionally fired wired-up charges prematurely, the lead-in wires acting as antennas. Certain safety procedures had to be worked out empirically to minimize this possibility, although by its very nature it could not be entirely eliminated.

Nevertheless, the average mine and miner remained faithful to Bickford fuse and to dynamite until well after the turn of the twentieth century. Miners knew how to handle both in reasonable safety, and as long as mining was conducted by relatively small organizations hoisting relatively small tonnages, there was no pressing need for any major change. The rise after 1920 of the very-low-grade-ore copper industry, blasting thousands of tons in one round from open-cast benches, put another complexion on matters, however. Once more research and change became the order of the day and doubtless will continue to do so for any foreseeable time to come.[20]

Chapter III

Amusements
and Diversions

AS late as 1935 every day was a work day in the mining camps.
Saturday still rolled around with fair persistence, and with each
alternate Saturday came the pay envelope. Tally-and-pay night produced
its quota of fights and sore heads lasting to or through the Sabbath—
"maze Monday" was the old Cornish expression for the day of an
unusually persistent hangover. Pay night also saw a rush of tramp
miners' business at the Stockade, Maiden Lane, or Floozie Barn, each
being a camp Casbah which respectable ladies professed to ignore but
which the children viewed with realism. Gunplay on Saturday night
seldom involved miners, being usually restricted to hard cases openly
"on the shoot," the horde of pimps, and the police officers who were
frequently recruited from the very ranks they were hired to keep in
an appearance of order. Miners wrestled for sport under Cornish rules
but settled grievances with their fists—being hit squarely by an en-
raged "bleddy fine double-'and 'ammersman," was roughly equivalent
to being kicked by a tramming mule. Taking one thing with another,
even such notoriously tough camps as Bodie, Creede, or Tombstone
were relatively demure in comparison to the Kansas cowtowns at the
height of the trail-herd season.

Alcoholic and amatory recreation aside, the inhabitants of isolated
camps were thrown back on their own resources when it came to seeking amusement. A great deal of this article was to be obtained from
the weekly newspaper, which usually arrived shortly after the first
saloon and brothel. The practitioners of this art form were as colorful
and boozy a set of men as the western frontier could boast. Given a
case of battered type, an arthritic Washington flatbed handpress and
some newsprint, they rested content in the pursuit and distribution,
if not synthesis, of news and beautiful letters. They would persist
therein as long as the ore body or the tolerance of the camp held out.

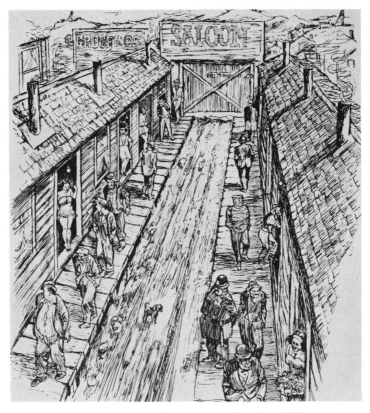

The stockade on pay night in the camp. The miners are looking over the girls, who are working in old-fashioned "cribs." Reproduced by permission of Buck O'Donnell and Shaft and Development Machines, Inc.

Some camp newspapers gained national reputations. Joe T. Goodman's *Territorial Enterprise*, of Virginia City, Nevada, was in the mid 1860's perhaps the most influential and best-written paper west of the Mississippi. The Tombstone *Epitaph* and its bitter rival, the *Nugget*, were also renowned. However, most of the camp papers sprang up like grass and usually withered as swiftly. Running simultaneously out of newsprint, advertising, and subscribers, the editor then loaded his equipment in a borrowed wagon and headed for the next mineral excitement with as much optimism as did the prospectors.

Anything not totally blasphemous or seditious was grist for the camp weekly. Its specialties were local stories—mine-development columns, police news, and social events (the last two categories often shading into one another). Mercantile and patent-medicine advertising revenue was important. Any remaining space was filled with extracts from other papers. Even such an influential professional journal as the *Mining and Scientific Press* of San Francisco found that its many mail subscribers demanded a few columns of world political news and human-interest stories distilled from the telegraphic reports. What passed for literature and poetry composed by the local bards and sagamen was included as well.[1] An editor like James W. E. "Lying Jim" Townsend, who haunted the High Sierra gold camps of Lundy and Bodie, found his tall tales, composed directly in type as they were conceived, usually well received if not well rewarded by his public. An editor could print what he pleased without much fear of libel suits, having virtually no property that could be levied upon to satisfy a court judgement. It was nevertheless understood that he might be required to give personal satisfaction executed on his body with horsewhip or pistol if he strayed too far from accepted editorial canons— James King of Williams was shot in San Francisco in 1855 for just such an alleged lapse of taste.

These editors and their reportorial and technical staffs (should they be able to hire any) were seldom far from a jug. H. L. Mencken spoke wryly of the "handset whiskey" favored by Baltimore compositors, and it cannot be supposed that the whisky improved the farther west one traveled. Townsend worked up a "cocktail route" in each of the camps he edified. It was the editors' pride that they got the paper out more or less on time regardless of aching heads, but on at least one occasion Townsend was obliged to reprint the previous week's issue with only the dateline reset, probably because his hands were shaking too badly to hold the composing stick.

In camps devoid of a press a Sunday amusement could be had by digging out a badger and setting the camp dogs on it. Procuring the badger was the most difficult part of the exhibition, since dogs were and remain a common commodity in mining camps. The dogs tend to be large, derived from a multitude of ethnic strains, and appear to be suffering from the last stages of *ennui*. During the day they lie about in abandoned attitudes, cluttering the sidewalks and doors of business establishments; only the appearance of a stray cat or a fight among themselves produces any collective signs of life. When the headworks' whistles blow to signal the end of the day, they arise, shake themselves briskly, and trot homeward to dinner, having put in a long shift. A badger, of course, arouses ambition in their breasts, but at Bodie a badger once whipped the dogs so handily that a street was named for him, and his exploits were recorded by no less an authority than J. Ross Browne.[2]

In later years the presence of a gullible greenhorn was likely to provoke the organization of a camp "badger fight" that required no preliminary digging. After being systematically alarmed by descriptions of the size and ferocity of the badger, the newcomer would be appointed timekeeper of the approaching match—a signal honor. Introduced to a crowded stable and confronted by a wooden box containing something alive and formidably resentful of its confinement, the dude would be handed the end of a rope that was said to upend the box so as to begin the combat. At this moment the cry would go up, "The badger has escaped!" followed by screams of panic and a mass stampede from the stable. Unconsciously clinging to the rope end, the terrified greenhorn would find himself a block or so down the street, alone except for a large chamber pot which he had been dragging behind him on the other end of the line. The sensibilities of the age being what they were, this experience would make the newcomer either conform to the camp's ways or depart on the first available transportation.[3]

One of the less appealing mining-camp recreations was the baiting of two species of comparatively helpless humanity, alcoholics and Chinese. There were a certain number of totally abandoned alcoholics in the West, forming a class which a later generation would call skidroad bums, who existed by working on construction gangs, mucking out corrals and stables, or swamping out saloons. For obvious reasons of safety, their presence was not tolerated on mining-company property. After one had achieved the temporary oblivion which he craved, sadistic practical jokers would have at him. On one occasion a number of comedians found such a drunk lying out in the sagebrush. They covered him with a long, narrow packing box of coffinlike proportions, and as he started to regain consciousness, they began to conduct a funeral service over the "remains." This ended in an abysmal fit of hysterics within the box, the practical jokers finding it funny. At the same camp a Chinese cook overtaken in drink (a peculiarity, for Chinese are as a rule abstemious) awoke to find his pigtail stapled to the wooden floor of the blacksmith shop on which he was taking his repose. He left camp, but his successor, another Celestial, took revenge by selling the kitchen beef to local Indians and feeding the crew on a stew concocted from the chipmunks which he had trapped.[4]

The "pasteboards" also helped kill time. Faro was the favorite western method of losing money in a hurry, and running a faro bank was a professional occupation, the dealer making an agreement with a saloon to provide him floor space and a table in return for an agreed-upon fee or percentage of the take. Besides the dealer-banker the game required a casekeeper to record the turn of the cards on an abacuslike board. Equipment included the casebox, or cardbox, and a rather elaborate felt table layout. The rules of faro, like those of Russian grammar, are incomprehensible to any mind not tutored to it from the cradle, but in essence the game resembled a sort of roulette, employing cards instead of a wheel and ball. Bets were placed on the card faces painted on the layout, and the action began as the dealer drew out the "soda," or valueless card, to expose the next in sequence, which determined wins and losses. Since the felt layout was usually decorated with the face of a grinning Bengal tiger, play became known as "bucking the tiger," while a heavy loser was said to have been "clawed by the tiger."

In the California placer diggings tedium was relieved by the founding of the Ancient and Honorable Order of E Clampus Vitus. The Clampers still persist as a dedicated but by no means sobersided organization of historically minded Californians, who devote much time and money to the commemoration of events and places which more formal historical societies are inclined to ignore.[5] It is reported that a most spectacular contribution in the manner of their spirit was the setting of a plaque, showing in relief a heart and a lighted lamp, in the sidewalk before a house which enjoyed the reputation as the longest continually occupied business establishment of its kind in northern California—this when the citizens of the community declined to observe or even admit the passing of a once-cherished local institution. The Clampers' marching song went somewhat as follows:

> Oh, what was your name in the 'States?
>   Was it Johnson or Williams or Bates?
>   Did you take a man's life, or another man's wife?
>   Oh, what was your name in the 'States?

All men were eligible for initiation into the Clampers, the fee being adjusted to suit the pocket of the neophyte and the thirst of the chapter. The ritual is described in some detail by an observer:

Lawyers, bankers, merchants and miners were members of this institution. And when the gewgaw, a big horn, rang out, for miles around miners came, stores and banks and places of business were quickly closed and all their managers soon repaired to the "Clampus Hall." The sounding of the gewgaw meant that a "sucker" had been caught and there were fun and beer ahead.

The candidate was prepared for the initiation by being divested of most of his clothing, then blindfolded. In this condition he was led around the hall, stopping at different points where he was catechised and lectured in a most fatherly way, by the different officers of the body. About the time he became worked up to the solemnity of the occasion, a strap with a ring attached, having been silently placed about his body, he would find himself suddenly lifted to the ceiling and then as suddenly dropped into a wheelbarrow, purposely prepared for his reception, in which had been placed large sponges saturated with ice water. The victim would be held thus securely in place while the wheel-barrow was run around for a hundred feet or more over a rough construction of round poles, jolting the wheelbarrow and keeping the victim bobbing up and down in a most ridiculous manner, on his ice-cold cushions. During this performance the members and spectators sang the while:

"Ain't you mighty glad to get out of the wilderness—

Get out of the wilderness,

Get out of the wilderness?"

Sometimes these initiation ceremonies extended over several hours. And by the time they got through with him the new member would feel certain that he had paid well for the entertainment of his friends; while he, himself, had added to his store of useful knowledge and experience.

Invariably the new member would steal out of town, humiliated and crestfallen, to appear again only when he could produce some new candidate or victim for admission to the order.[6]

Mining-camp celebrations of the very first chop fell on Christmas, the three days of the Fourth of July observance, and certain movable feasts which were sponsored by any prospector who had just cashed a large check in payment for a high-grade location.[7] The patriotic festival emphasized the discharge of explosives, an activity familiar to every miner and requiring materials very easy to obtain. Dawn was saluted with a major blast followed periodically by others, and the intervals were filled with the rattle and snap of pistols being fired at the sky. The three days were devoted to parades and contests of skill and were usually concluded with dances at the union or lodge halls.

One of the unique aspects of camp Fourth of July celebrations was anvil shooting, or anvil salutes, an old frontier custom slightly modified to minimize one possibility for permanent personal embarrassment. Originally a blacksmith's anvil was dragged from his shop into the open air. The square holes in the heel were filled with black powder, and a fuse or powder train was laid to the back edge of the heel. A second anvil was then upended and lowered atop the first so that their upper surfaces met and matched. The train was lit with a long portfire, and in a moment there followed a ringing blast and a cloud of white smoke out of which the upper anvil would sail grandly, end over end, through the air. This was all very well but for one contingency. When fired in this manner, the upper anvil's heel would be violently elevated first, thus tilting it and dashing its curved horn against the corresponding horn of the lower anvil. Often enough this was more than the cast iron could stand, and one horn or the other would be broken off. Such a dehorned anvil was said to have been "muleyed" and was considered a standing reproach to its owner, who had evidently got too drunk to withstand this abuse of his shop's chief ornament.

In the western camps the ingenious miners worked out an improvement on the process. Instead of placing the powder in the heel holes, they prepared the anvil in a radically different manner. A wagon-wheel hub nut was procured, and a deep groove was filed into the rim on one side, the groove extending from the inner to the outer edges. An inch of dynamite was cut off the stick, primed with detonator cap and Bickford fuse, and set in the nut, with the fuse rattail led out through the filed groove. This loaded torus was then placed in the exact center of the lower anvil's surface and the upper anvil balanced on top of it. The fuse was split and fired, but since the charge went off directly beneath the top anvil's center of gravity, the anvil flew up, keeping its wings as it were level and true. It was hoped that it would descend slightly to one side or the other, for if in its fall it struck the launching pad, another dehorning might well result.[8]

The Fourth of July Grand Ball was attended by everyone in camp who was still on, or approximately on, his feet by nightfall. Since by unspoken custom it was barred to the "fair Cyprians," camp wives would attend freely, and rancher's families came in from as much as a hundred miles. Younger matrons (it was more common than not for a girl to be married at the age of fifteen) of necessity brought along their babies, for whom a nearby nursery was provided. Once the babies were well asleep, their mothers departed eagerly for the dance floor. This gave practical jokers an opportunity to move in. Waiting until the mothers were engrossed, the comedians would steal into the nursery and quietly shift the babies about, making certain that a given child's wrappings were left in the original location. When the dance broke up, the mothers would slip into the darkened room, lift a child from the place where she had deposited her own, and hurry home. With the morning the exchange would be discovered, and, after some moments of hysteria in each household, the mothers would go forth to swap babies back and forth until all were restored to the bosoms of their families.[9]

The days of the celebration were given over to gentlemanly contests of skill, with something for everybody. The union and the lodges provided brass bands and perhaps fielded a baseball team. The volunteer fire departments had hose-cart races. There might be a special ball game of engineers versus miners or bachelors pitted against the married men. At Bodie the benedicts were encouraged to hit by someone who set a keg of beer at first base.[10] Foot races and tug-of-war contests were sponsored. Children ran riot and were surreptitiously encouraged to gorge themselves into fits of vomiting. The ladies looked forward eagerly to the dances, since they were always so outnumbered that every woman, be she ever so grim or angular, could count on more invitations than she had lines on her card. Possibly the only people not to relish the occasion with whole heart were the camp physicians and the board of education. The physicians would have work enough patching up the enthusiasts afterward. The board was sadly aware that the dances would play havoc with the coming year's faculty arrangements. One mining executive noted:

It has been said that one serious problem is to keep down the employment turnover among the young lady teachers. There always is a considerable contingent of young engineers on the mining company staff and invariably a dearth of young women of marriageable age in one of these (mining) communities. The probable fate of a personable young schoolma'am is easy to imagine.[11]

The afternoon of the Fourth itself was devoted to serious business— the drilling and mucking contests. The entrants competed for purses made up by civic-minded merchants and displayed their talents before large crowds assembled around the elevated boxing ringlike platform erected in the center of the camp. Such contests unquestionably had originated informally as a result of lunch-hour boasting in a drift-level station or in pay-night saloon challenges, but they quickly moved into public view. Since they had every qualification to succeed as spectator sports, the rules became institutionalized, and the participants ultimately approached the training standards and mental outlook of modern sports professionals.

The mucking contest was simplest of all. A one-ton ore car was lifted to the surface of the heavily built platform. On the ground below its forward end, about eight feet below the car rim, was a large plank bunker, or "sollar," containing one ton of assorted sizes of rock waste. The object of the exercise was for the contestant to transfer the contents of the pan upward into the ore car in the shortest possible time, using his hands for the larger pieces and a number four scoop for the remainder. Timekeepers checked the performance and noted elapsed time between the firing of the starting pistol and the moment the last scoopful of muck—and the scoop itself—sailed into the car. The time was chalked next to the contestant's name on the scoreboard, and the car was dumped "end-o" to spill its contents back into the pan in readiness for the next contestant.[12]

A micromucking contest prevails to this day in the national gold-panning championships. Counted numbers of gold colors, usually sixteen, are placed in the standard eighteen-inch pan and stirred up with a shovelful of silt, gravel, and pebbles. A waist-high cistern of water is provided, and at the starting signal the panner begins to toss away gravel, swirl the pan, and otherwise make the sand and water fly. When the sixteen colors are neatly aligned in a clean tail along the bottom edge, the stopwatches are halted. Penalties are assigned for the loss of any of the sixteen colors, but a real expert can clear the pan and yet retain every color in well under one minute.

After the mucking had warmed up the crowd, it was time for the drilling contests. The contest was conducted on a large block of carefully selected Vermont or Gunnison granite at least six feet thick, the ample upper surface of which had been dressed flat and set nearly flush with the top of the platform. This block had been quarried and shipped in at considerable expense, and the block itself was carefully inspected for uniformity of structure and absence of flaws or cracks. Granite was preferred not only for its hardness but also for its freedom from vice. Sedimentary rock was too soft to offer much challenge, basalts were far too tough, quartzite was too abrasive, and so forth. The drillers themselves had practiced for months, and they lavished enormous care on the forging, sharpening, and tempering of the drill steel which they themselves provided, each according to his own theories of effectiveness. Having a steel jammed, or "fitchered," in an invisible flaw of the block was regarded as hard luck, but a man whose steel broke in the hole as the result of improper tempering had only himself to blame for a misfortune which would surely put him out of the money—a broken bit either jammed the hole beyond redemption or had to be drilled through by another steel at great cost in effort and fatal loss of time.

At the height of the drilling contests competitors were divided into heavyweight and lightweight classes, and the events in each class were single-jacking; team, or "change," double-jacking; and "straightaway" double-jacking. In the first a single contestant drilled by himself, holding the steel in one hand and smiting with a four-pound hammer in the other. Ambidexterity was a great help, since he could shift the hammer from one hand to another so as to distribute the work. Nor did he merely pound away like a carpenter, for the single-jack hammer was equipped with a wrist thong like a cavalryman's sword knot. On the backstroke the fingers were opened and relaxed, allowing the han-

Single-jacking. The miner is opening his hand on the backswing, allowing it to relax, while it is retained by the wrist thong. Reproduced by permission of Buck O'Donnell and Shaft and Development Machines, Inc.

dle of the jack to swing free, restrained only by the thong. At the end of the backswing, the fingers again closed around the handle of the jack to bring it down for the next stroke. At frequent intervals the other hand drew the dulling steel from the hole, flipped it aside, and dexterously inserted a new, slightly longer steel during the unbroken sequence of blows. Single-jacking was done on top of the block with the driller kneeling.

The downholes had to be cleaned of the rock cuttings which accumulated therein as the steel chipped downward into the granite. In conventional mining such cleaning might be done with the miner's copper spoon. Ordinarily, however, drift holes were kept clean by the much simpler method of pouring or dashing a splash of water into them from a tomato can. When done properly, the water would combine with the cuttings to form a tenacious mud, which adhered to the steel and could be withdrawn, clinging to it. Rapping the steel sharply against a rock then knocked off the mud. In contests a hose of running water was usually provided, tended by a nonpartisan official who expertly added just enough so that the pumping action of the steel splashed out the water and removed the cuttings as fast as they were generated.

These events were fifteen-minute affairs during which every second counted and no unnecessary moves were made. The double-jacking change contests involved two-man teams. One man held and rotated the steel, while the other pounded away with a double-jack whose weight was limited to nine pounds (handle included). The partners exchanged positions and steel at one-minute intervals. In the straightaway, one man did all the drilling, his partner "turning" for him. Instead of the ordinary drift pace of fifty blows a minute, the contest pace was forced as high as eighty, and, everything else being equal, the team which could press the pace was likely to be the winner. They appeared on the platform in shirtsleeves, carrying a bundle of fifteen steel which they had nursed like babies through the smithy. Each steel was destined to be used but one minute and for 2½ or 3 inches of hole, whereas the ordinary drift steel was expected to last for 6 inches and five minutes before it was considered too dull to go further. At the end of the minute the steel was pulled and tossed aside and another gracefully substituted with never a break in the steady ringing of the double-jack.

These steel varied from the short, stubby "starter," which was used to begin the hole, to the specially forged five-foot steel, which, they hoped, to employ at maximum depth during the final minute.[13] Rules

Ross Thomas (left) in a straightaway drilling contest, Telluride, Colorado, July 4, 1908. Note the steel flying from the holder's left hand. Courtesy Ross Thomas.

limited the double-jack steel to a 7/8-inch and the single-jack steel to a 3/4-inch diameter. Aside from the fine gradations of length, they appeared to be identical, but every miner in the crowd knew that the chisel-bit width from ear to ear not only was narrower than usual[14] but diminished in minute increments with each increase in over-all length so that the bit would "follow" its predecessor easily in the hole without jamming. Proper temper and sharpness of the bits were of intense concern: a badly tempered steel might snap an inch above the flare (the depth to which the smith submerged it in his tank[15]), or, not quite so disastrously, one ear might break off if the curvature and temper of the cutting edge were not absolutely true. Though there was no earthly way of fishing out the metal, custom demanded that the team insert the next change and continue drilling, even though they were foredoomed to finish out of the money.

There could be even worse things. The Tonopah and Goldfield camps in Nevada were at the turn of the twentieth century hotbeds of a great many activities, including drilling contests. A young bride, Mrs. Hugh Brown, witnessing one contest, commented:

A drilling contest has everything: technique, beauty, endurance, speed and danger. If the hammer descends a fraction of an inch out of line on the tiny head of the drill, a man's hand may be crushed.

Once during my life in Tonopah, I saw a man's hand struck. Suddenly the hammer poised in midair. The crowd groaned, knowing what had happened. After an instant flinch, the man crouched over the drill looked up at his towering partner, and yelled, "Come down, you!" Down came the hammer. The men cheered and the women cried. The hand on the drill began to turn red, but still it held on to the drill. When the injured man's turn came to rise and hold the hammer, the blood crept down his arm until it looked as if it had been thrust into a pot of red paint. The blood ran into the hole and mixed with the water from the hose. Everytime the hammer descended, the red fluid sloshed up and spattered nearby onlookers. The man sagged lower after every blow, but he never gave up until the timer's hand signalled fifteen minutes. Then he fell over in a dead faint. The platform looked like a slaughtering block.[16]

After the halt was signaled, the last steel was pulled, and the panting drillers relaxed while a steel measuring rod was inserted into the hole. A sliding ring clamp was slipped down to touch the surface of the block, tightened, the length carefully measured, with the results marked on the scoreboard. Nevertheless the audience and participants already had a fair idea of the outcome since it was the custom to count and announce the number of blows struck during each successive minute, both to help the contestants pace themselves and for the convenience of the onlookers. Unless a driller was coming down lightly, the faster was usually the deeper.[17]

When the winner in each event was announced, a roar went up. The rest of the afternoon was then devoted to purse awards (as much as one thousand dollars), champagne, and the collection of bets, since the contest drillers had no hesitancy in placing side bets on themselves. For that matter, hustling by experts in remote camps was not unknown. A pair of ten-day miners would drift in to seek work and, while giving every appearance of naïveté, would begin to boast of their drilling prowess in an irritating and unconvincing manner. When the local champions learned of this, a contest and bets would follow immediately. At the appointed hour disillusionment would begin to set in, starting with the moment when the new arrivals produced their expertly sized, sharpened, and tempered bundle of contest steel. The measurement of the holes and collection of money afterward was little more than a formality.[18]

Perhaps as early as 1885 district champions started to flock to national eliminations held on the Fourth of July at such big camps as Butte, Kellogg, Tonopah, and Bisbee. The national winners received as much adulation as toreros in Spain. "Champion drillers were kings, known and feted throughout the mining world. The prize money was accompanied by cases of champagne and other liquors,"[19] a hint that total abstinence was not considered essential to training standards. Some of the records are worth repeating. At Tonopah in 1905, Fred Yockey, of Colorado, won $587.50 for drilling 27 7/8 inches in the single-jack event. Earlier, at Bisbee, Arizona, in 1903, Sell Tarr, "the King of Drillers," had put down 28 5/8 inches in the straightaway contest.[20] In the Mining Congress at El Paso, Texas, also in 1903, the team of Chamberlain and Carl Make drilled 42 1/2 inches in fourteen minutes. They deliberately loafed during the last minute, partly so as not to discourage future betters and partly because drilling completely through the block constituted disqualification. In this contest they were assisted by an old Swedish hose holder, who added beer to the water on the theory that the foam would bring the cuttings to the surface faster. His advice on pacing stands as a monument to coaching of all sorts, being, "Now get oop and hit him yust som hard and yust som fast som you can."[21]

The camp Christmas was an occasion for mixed emotions. Families observed the holiday in seclusion at home, with the Irishry turning out in great numbers for midnight mass. For the drifters, tramp miners, and habitués of the licensed quarter it was a time of sadness and sentiment, which they attempted to alleviate with alcohol, often enough only making matters worse. Christmas Eve was the one time in which the resident faculties of the cribs and floozie barns were permitted out of bounds. They congregated at the saloons, where some pitiable attempt had been made to deck the hall with red paper and mistletoe and to decorate the forlorn sagebrush which had been dragged in from the desert to do duty as a Christmas tree. The "Professor" played carols on the piano, while the celebrants made a mighty effort to pretend cheer, the prostitutes exchanging presents with each other and bestowing expensive gifts upon their pimps. The male portion of the crowd affected an untoward decorum, knowing that the girls never discussed or did business on Christmas Eve.

Despite all efforts the emotional climate of the evening invariably went from blue to black. The camp physician kept the harness on his stabled horse and went to bed fully dressed, knowing from experience that shortly after midnight he would be summoned to the bedside of a depressed prostitute who had drunk carbolic acid or some other corrosive disinfectant. When he arrived at his destination, there was little enough for him to do but get out the morphine and go to work. The madam took matters in hand from there, knowing that the girl would wish to be buried in a white wedding gown, with services attended only by her close friends. Neither the camp newspapers nor the memoirists as a rule have a great deal to say about Christmas—life on the mineral frontier was not always kind.

Other Christmas trees would be set up in the sitting rooms of the boardinghouses, regular institutions in every camp of moderate size and longevity. These establishments were as standardized as military posts, devoted to grim propriety, and as caste-bound as Hindustan. Fire and desuetude have destroyed most of these hostelries, but one

The W H Hotel, Mayer, Arizona.

is still to be found doing business in the once-prosperous Mayer, Arizona, camp. Over the veranda steps hangs a modest sign proclaiming it to be the W H Hotel. It is evident that the management soon had second thoughts about the message, for below it in rather smaller letters the words *White House* were added by way of exegesis. Set within grounds of ample extent, the hotel is of two-story brick construction, topped by a sloping tin roof. Its cubical lines are relieved by the two-story wooden veranda extending along the south facade and the west and back sides. On hot summer nights this airy arrangement served as a lounging and sleeping area onto which roomers could drag their mattresses when the temperature in their poorly ventilated rooms proved too much to bear.

Immediately inside the double-leaved front door is a lobby or lounge of moderate dimensions, separated by a wide arcade from the larger dining room behind it. The room is capable of holding two long tables accommodating twelve places at each table. At the left of the sitting room is the former parlor and office of the owners, and directly behind it, opening into the dining room, is their bedroom. Taking up the full width of the first floor rear is a large kitchen, complete with pantry, long wood-burning range, and a big walk-in icebox. The visitor notes that the head of the owners' bed is against the kitchen wall, making it difficult for a midnight prowler to forage for a snack without detection.

Stairs lead up from the left side of the lobby to the second-story hall, which runs to the rear of the house. On each side off this hall are three single rooms. Behind the stair landing at the front of the house is a suite of two larger rooms which share a separating but private foyer. Such grandees as territorial governors, bishops, and financial magnates were evidently accommodated here, whereas lesser lights lived in the rooms off the hall. All these guests were seated for meals at the dignitaries' table headed by the owner of the hotel. Toilet facilities in all rooms were minimal, being limited to pitcher, bowl, and a chamber pot discreetly slipped beneath the bed for convenience in the night or wintertime. Otherwise, two structures of simple architecture were to be found in the backyard by those who would seek them. Ladies, apart from newly arrived brides of salaried personnel, very seldom lived at these hotels, and those who did departed as soon as private quarters could be found for them in the camp. Other women were not accommodated.

Fifty yards behind the main house are a number of two-room bunkhouses for working miners. They lived four to a room and ate at the second table in the dining room, to which all were summoned by a lusty beating of an iron triangle which hung from the veranda. Since the bunkhouse residents daily showered at the mine's change room when going off shift, they needed only a communal shaving pitcher, bowl, and mirror. Though they shared the same fare as the upstairs guests, they probably paid a dollar or so less a day for room and board. The bunkhouse lines were extended to the east (the lee of the prevailing wind) by the stables, in which the guests' animals and vehicles were housed and cared for.

The staff of such an establishment consisted usually of the owners, husband and wife; a cook, either a male Chinese or an Irish woman; two maids of all work; and a handyman, who also served as the hostler. The maids made up the rooms, washed dishes, waited on tables, cleaned, laundered, and polished, but their recruitment, training, and retention were usually a constant source of anxiety to the lady of the house in whose charge they were. Ranch girls were preferred, since they were accustomed to such work and were glad to be where their chances of finding a husband were far better than in the isolation of their homes. Otherwise—and many Mormon ranchers did not care to have their daughters exposed to gentile influences—a couple of colleens might be enlisted. They would not stay long, either, since the priest would soon be after finding them a Patrick Francis or a Michael Joseph to call their own. Should worst come to worst, the mistress might have to coax in, scrub, and semicivilize a pair of aboriginal lasses, but of her difficulties in this department the less said the better. In all cases, however, the maids' virtue was rigorously guarded, and no familiarities by either the quarterdeck or the fo'c'sle were tolerated.

The boarding-hotel chaperonage was of such high voltage that the hotel lobby or the veranda was nearly the only place in the camp where a young engineer could take a pretty schoolma'am for an evening's entertainment without giving the camp gossips an infinity of mill grist. And with the grievous shortage of marriageable young women, most courting was of necessity entirely honorable and above board. Such amusements were available as veranda conversation, uplift committee endeavor, genteel singing, and the recitation of home-brewed poetry. Mining men as a class were fond of rolling their own metrical sagas, while the manufacture of sentimental verse was esteemed to be a ladylike accomplishment. Fortunately for American letters, most of this camp verse has vanished. Robert W. Service was regarded as the Vergil of the camps, although his strophes were for the most part considered too realistic for a lady's ears.

In addition to its other functions, the hotel was the nerve center of the area, providing space for political and financial assignations of all sorts. In its hall rooms consulting engineers wrote confidential reports and wrestled with technical puzzles. In the grand suite politicians prepared their speeches, held levees, and distributed patronage. The need for privacy was greater than the desire for ventilation; hence there were no transoms over the doors, which themselves were of uncommon thickness. Since there was no more central heating than there was ventilation, each room was warmed by its own miniature cast-iron base-burner stove, little more than two feet high, fed with bits of mesquite wood fetched around by the handyman. The strategic location of the proprietors' bedroom enabled them to detect any unseemly noises from above, though not to eavesdrop on low-voiced conversation.

The furnishings of most of these boarding hotels were replaced many times over, yet the probability is great that the pieces originally on public display and in the front suite were massive, ornate, and unsparing of brass and gingerbread carving. The decor declined with the social scale, and in the miners' bunkhouse was spartan. Where the hired help slept is difficult to imagine; there was no basement, and the attic air space was uninhabitable. There may have been a curtained alcove in the kitchen to accommodate the maids. The masculine employees probably slept in the bunkhouses or the stable harness room— it would be well in fact if they did sleep in the stable, since they would be present to smell smoke and to drive out the stock at once, should fire break out during the night.[22]

Not all miners lived in such grand surroundings as these. In fact, the great majority of bachelors sought room and board in the house of a fellow worker, the Irish living with the Irish and the Cornishmen with Cousin Jacks. The three or four dollars a week that a boarder contributed made a welcome addition to the family income, but it was understood that the boarder worked the same shift as the head of the house. Sociologists have not dwelled on the fact, but the bachelor boarder's constant exposure to the needs and requirements of the many children always underfoot probably constituted a valuable basic training for the day in which he himself would assume the duties of a paterfamilias—or, in any event, somewhat cushion the shock.

These miners' homes were very modest frame structures, so inflammable that the inevitable big fire which sooner or later visited every camp has wiped out nearly every one. The women did their best with curtains, a pot of geraniums, and bright lithographs to improve the interiors, but the yard and street were usually hopeless tracts of sand or alkali dust, once in a great while sprinkled by the city watering cart. Indeed, the remains of such camps as Virginia City, Goldfield, and the like are mostly composed of the stone or brick business-district buildings, thus giving the visitor no good idea of their much greater extent during the flush times when most of the population lived under wood or even canvas the year around.

Even modern mining camps are seldom the loveliest villages of the plains, though this is due chiefly to the scarcity of water. Company towns used to prefer concrete business and office buildings which when well designed might be attractive but if otherwise were lamentable. The private housing was depressingly uniform and was made worse by the habit of painting every exterior wooden surface a dark, dingy green which was as characteristic of company property as rusty maroon was typical of the railroads. Privately owned homes were seldom painted at all, but certainly were never white—the only white pigment then available was based on lead, which the sulphurous stack gas of the smelter would quickly reduce to an unspeakable dark gray.

Every camp had a full share of boarding houses operated by widows. The hazards inseparable from underground work contributed heavily to this situation, the greatest killer of all being silicosis, or miner's consumption. Gold, silver, and copper mining are almost always conducted in highly silicious rock, and virtually every step of the process creates (in the absence of preventive measures) huge clouds of dust. It did not require much exposure to such conditions to carry a miner off rapidly. Three centuries previously Georg Bauer had noted that a high death rate went along with hard rock mining, saying "[The dust] eats away the lungs and implants consumption in the body; hence in the mines of the Carpathian Mountains women are found who have married seven husbands, all of whom this terrible consumption has carried off to a premature death."[23]

A widow could do little else to feed herself and her children but to capitalize on her only skill, housekeeping, by taking in two or three boarders. It would be inevitable that her daily exposure to a number of single men would soon bear fruit in the admission that life must go on, leading to remarriage and a new crop of children. Despite the human wastage, however, the western miners and their families did not appear to display the apathy, pessimism, and downright hatred for their work that seems to prevail in underground coal mining. "Pit is a devil," said

Welshmen of their collieries, and the coal regions of American Appalachia are notorious. But this did not hold true for the optimistic, proud, and professionally oriented western miner. Part of his attitude may have stemmed from frontier optimism and individualism, but a good part can be traced to the vocation itself. "[Hard rock] mining," concluded Bauer, when all was said and done, "is a calling of peculiar dignity," and the Cousin Jack agreed wholeheartedly with him.

While on the job miners had their own professional amusements and practical jokes, a selected bouquet of which is illuminating. These *divertissements* were most often performed during the half-hour "croust," or lunch break, although no rule restricted time and occasion. For instance a pipe tender — a sort of itinerant inspector and repairman of water and compressed-air piping—working in the Homestake Gold Mine was fastened upon, early in one shift, by a stope crew. He was spreadeagled against a timber wall and nailed securely into place with drift pins. Unharmed but helpless to move or free himself, the pipe tender had to remain where he was for some time before a passing shift boss noticed him and freed him from his predicament. Another man was stripped naked, forced into a cage, and the cage "rung off," or hoisted between levels, to be left idle. The miner, having no means of signaling, was left in the cold, dark, and damp for fifteen uncomfortable minutes before the cage was in movement again.

In mines such as Bisbee, where mule tramming was employed, the natural residue of these animals was employed in many ways not mentioned in mining textbooks. A man's jumper, left hung on a nail in the level station, was an invitation to have its pockets rammed full of the ecological by-product. Mule droppings could be flung, used to bombard men working at the foot of raises, and otherwise adapted to uses limited only by ingenuity. The standard favorite was to remove the pasty from a lunch bucket and refill the compartment with road apples and often nail the lunch bucket to a plank as well.

When the one-dollar Ingersoll pocket watch became available, these cheap timepieces instantly gained favor with miners, who carried them in the jumper pocket with a shoelace "chain." They kept time as satisfactorily as any expensive railroad Hamilton and when inevitably mashed, ruined by damp, or simply lost represented no economic calamity. But when a man had lost or ruined several Ingersolls in quick succession and then lamented too audibly about it, the crewmen would bide their time until he brought down a new one. As soon as possible it would be lifted from his jumper and displayed in the station, spiked up with a nail driven through the center of the face. The owner's attension would be called to it with the remark that "here was one he couldn't lose."

More horseplay at croust time involved the common practice of catching a short nap on a plank or in a wheelbarrow. As soon as a man began to snore, comedians would advance on him and quietly fire the ends of his Bickford-fuse belt string and pants "Yorks." The burning fuse warming his waist and upper calves would soon stimulate the sleeper into a credible imitation of a man who had put his foot into a hornets' nest. A variant of rousing the sleeper was played on a miner who constructed a crude hammock for himself at the bottom of a raise. Resenting this luxuriousness, the crew working above simply opened a four-inch waterline and flushed him out of his nest.

In a later and more effete period the good old dirt-filled candle box which served as a privy in the level station was replaced by the "red wagon." This was a two- or three-hole portable Chic Sale or potty car, dragged to work caboose-fashion at the end of a man train, spotted in a central location, then hoisted to surface now and then for emptying. (A well-preserved example of such a car is in the Mining Museum at Jerome, Arizona.) The red wagon's strategic possibilities instantly commended themselves to honest miners everywhere. At the Homestake, for example, one crew fell into the habit of blasting early, allowing the smoke to go into a nearby stope to the great annoyance and discomfort of the crew in labor there. At last the offended crew applied a scoopful of "eau de red wagon" to the smoke-producers' vent fan. This produced an instant truce.

Finally, the same Homestake car occasionally lent itself to another prank. The car in question was spotted on a short spur track coming off an inclined haulage line on the 3,050-foot level. From this point the line had a considerable grade down to the hoist shaft. It was economical in moving loaded ore cars, since gravity eased the work considerably, but anything on wheels would roll swiftly down the grade. Catching a fellow worker in meditation on the red wagon, the practical joker would quietly set the track switch and give the car a strong push.

Before the contemplative could realize what was afoot, he and the car would be out on the grade and accelerating. With his ankles entangled in his trousers, the victim could do nothing but hold on grimly for a wild half-mile ride, ending in a swift and loathsome manner at the hoist-shaft station.[24]

Amusement was also had by parodying the instructions of shifters and level bosses, the following humorous decalogue yet containing some substantial blocks of advice and admonition:

I

Thou shalt not slumber late in the morning, but shall rise ere it is day and break thy fast, for he that goeth to the mine late getteth no candles, causing the transgressor to grope in darkness and the shift-boss to indulge in profanity.

II

Thou shalt not take up thy position in the center of the cage when descending or ascending the shaft, neither shall thou appropriate in thy person more room than the law allows, for thou are but of little consequence among a whole cage-load of men, no matter what thou thinkest to the contrary.

III

Thou shalt not hesitate on the station, or smoke thy pipe and talk politics with the pumpman, for verily the shift-boss might suddenly appear, and heaven help thee if he findeth the (ore) chutes empty.

IV

Thou shalt not mix waste with the ore, neither shalt thou mix ore with the waste, thou nor thy partner, nor the mucker within thy drift, for surely as thou doest these things the mine will stop paying dividends, and thy name will be "mud" over the length and breadth of the camp.

V

Thou shalt not eat onions when going on shift, even though they be as cheap as real estate in Clifton [Nevada], unless thy partner participateth likewise, for that bulbous root exciteth hard feelings in the heart of the total abstainer, and causeth the interior of a mine to be an unpleasant place.

VI

Thou shalt not address the boss by his Christian name, neither shalt thou contradict him when thou knowest he is lying, but thou shalt meekly say "Yes" or "No" to all that he suggests; and laugh when he laughs and keep on laughing when he relateth a story, even though it be older than thy grandmother.

VII

Thou shalt not steal thy neighbor's mops,[25] nor his picks nor his drills; neither shall thou carry away on thy person or in thy lunch-bucket low-grade ore from the mine,[26] for thou wilt find it will take a lifetime to obtain a millrun.

VIII

Thou shalt not have an opinion concerning thy place of work, for thy employer payeth a fat salary to a school-of-mines expert for constructing in his mind bonanzas that don't exist, so thou shalt refrain from theorizing, and concentrate thy efforts on drilling and the blasting of an abundance of powder.

IX

Thou shalt not, in order to breathe, steal from the drilling machine compressed air intended for drilling purposes. Thou shalt not go on strike lest thou be turned adrift on a cold and cheerless world; neither shalt thou demand thy pay, for the company's directors in the East know not that thou liveth, neither care they a tinker's dam.

X

Thou shalt work and break ore every day, the Sabbath included, for verily the board of directors aforementioned hath assumed the prerogatives of the Almighty, and if thou refuseth to toil as they dictate thou and thy dog and all that thou possesseth will soon be hitting the trail for Tonopah.[27]

CHAPTER IV

Silver Prospects

F RAY Marcos de Niza and after him Francisco Vásquez de Coronado marched north from New Spain in the 1530's to search the Back of Beyond for golden cities half rumored and half hoped for. Finding nothing which resembled their goal, the Spaniards retreated to the southern margins of the great Sonoran Desert for a generation. Tentative probings by later explorers, such as Antonio de Espejo and Fray Francisco Sylvestre Escalante, suggested that the mountains of the North had some silver mineralization but that it was in no way comparable to the enormous silver lodes developed about Mexico City after 1550. Over the course of the next two hundred years, however, Hispanic influence gradually edged northeast to the Gila River and the upper Río Grande. Silver-bearing outcrops of moderate tenor were worked in the Sawatch Rockies and also near by the wretched hamlet of Tucson and the Santa Rita Mountains. Distance from Mexico, desert, and the ferocity of the Indians made these developments amount to little more than intermittent chloriding.

The surge of Americans into the Southwest after the Mexican War changed the situation considerably. The adventurers Charles D. Poston and Sylvester Mowry not only worked the Santa Rita silver lodes but greatly magnified their exploits in print. The California gold rush created an army of prospectors, who fanned out in search of mineral deposits. Although many were called by the gold bug, few were chosen by good luck. Despite the known odds against success, these prospectors greatly enjoyed their nomadic, venturesome life, and, as befitted professionals, they left as little as they could to chance. A prospecting expedition was financed by undesirable but essential toil as a quartz-mill hand, miner, carpenter, or teamster. For those men resolutely opposed to compromising their principles, there was hope of getting a grubstake by persuading a friendly storekeeper to advance a month's food and the loan of a pack animal. The capitalist accepted in return a verbal or written promise of a half share in anything which might be located. So equipped with the necessities of life and labor, the desert rat went forth to conduct his search in a logical and systematic manner within his capabilities.

The experienced prospector commenced by hunting up a district which was known to be mineralized or which looked to him as though it might carry mineral values. Years of apprenticeship, experience, and campfire discussion suggested what such country looked like. To sum up a very large body of "signposts" and lore, it may be said that promising country was usually rugged and that many of its rock formations were quite colorful as a result of iron stain—whole mountain peaks of the Sawatch Rockies are brilliant red or yellow above the timberline for just this reason. It was a favorable sign if a great deal of quartz was to be found in the stream gravels and outcropping on the hillsides; white quartz was good, but if it was cindery and very "rusty," that was even better. Tangled, confused granitic and schistose country much penetrated by quartz threadlets and stringers was also a very good indication, particularly if the rock had a brown-black "burned" appearance.

The veteran made his way toward such major formations and began checking them by panning the gravel of the streams and washes at their feet for gold or other heavy minerals. He could knock chips off the weathered and decomposed outcrops, crushing the chips in his field mortar (the sawed-off bottom third of an iron mercury flask) and panning this dust for values. All the while he kept an eye open for "float," the name assigned eroded rock fragments which contained mineralization or appeared as though they had once done so. When a prospector spoke of "rich float," he usually meant rusty-colored spongy quartz, which close inspection would show penetrated by cavities which were perfectly square.[1] Such cavities had once held iron pyrites (fool's gold), one of whose crystal forms is cubic. When exposed to atmospheric weathering and surface water, the pyritic mineral was decomposed much sooner than the quartz which included it, leaving the telltale cavities. Iron-, copper-, and other pyritic-base minerals are worthless in less than hundred-ton lots delivered to the smelter yard, but at depth they often carry substantial gold values. Gold will not be decomposed by weathering, and often enough rusty quartz in place contained valuable bits of native gold in the cubical voids, gold which could be recovered by crushing and panning outcrop specimens or by washing the detritus at the base of the formation.

Silver, usually accompanied by variable percentages of lead, represented a somewhat different problem to the prospector. The two metals usually appeared together, many times associated as well with zinc, and at depth were chemically combined with sulphur. These sulphides were called not pyrites but silver sulphurets, galena, and zinc blende, respectively. Iron seldom accompanied them in such quantity as to stain their gangue rock a decided rusty maroon, but there was usually enough iron to contribute a light-brown outcrop stain. Silver sulphurets and galena weather down by stages, the first of which appears as dull black or brownish red. They next may be further reduced to "horn silver" of light green or beige hue or to lead carbonate ranging in color from light yellow to a fluffy matte white. In one final stage the atmosphere reduces them to silver-lead argentite, black, readily flattened nuggets of great mass and metallic appearance. A great many silver-lead deposits have therefore been first located by the discovery of such black nuggets at the surface of the lode outcrop.

Such a discovery is well illustrated by the history of the once-famous Silver King Mine, observable a few miles distant from the present copper camp of Superior, Arizona. In 1871 this region was filled with warlike Apache Indians. To alleviate the danger, the area commander, General George Stoneman, had established a small military outpost, Camp Pinal, at the point where the Apaches' favorite war trail debouched from the Pinal Mountains into the low-lying desert on the south. A short while later, with some idea of permitting cavalry detachments to chase the Indians back up into the mountains, Stoneman decided to improve their trail. Army fatigue details were set to work on "Stoneman's grade," which wound up from the desert, passed through a spectacular natural amphitheater, and so into the Pinals.

Among the troopers who labored with pick and shovel was a soldier named Sullivan. Returning one evening from the road-building detail, he paused to rest on a small, light-brown conical hillock centered in the great semicircular valley. Sullivan knew nothing about mining, which was a pity; had a Mexican *gambusino* seen that hillock, he would have emitted yips of glee, for it was a miniature version in every respect of those conical brown hills of metropolitan Mexico within which the greatest silver lodes of all time had been discovered three hundred years before. Idling as a soldier will, Sullivan found some heavy, black pellets which, to his surprise, flattened easily when pounded between two rocks. Thinking little more of the matter, he pocketed a few and went his way back to the Pinal picket post.

When Sullivan's enlistment was up, he drifted south to Charles G. Mason's ranch on the Gila River, where he showed his nuggets about but declined to reveal their source. After a time Sullivan departed and was no more heard from; in thinly populated Arizona, where everybody knew everyone else's business, it was automatically assumed that Sullivan's going missing meant that the Apaches had killed him. The memory of his nuggets persisted, however, and a number of prospecting parties scouted the Pinal Mountains for their source, but without luck. It is a moral certainty that these parties and perhaps dozens of other people passed for months on end the hillock at the foot of the grade but never paused to investigate it.

In 1874 the great copper deposit at Globe, Arizona, was located high in the Pinals and not far as the crow flies from Sullivan's hillock. The Globe strike was a lode of high-tenor, direct-smelting ore, good enough to repay the costs of packing out by way of Stoneman's grade. The rancher Mason, with partners Benjamin W. Regan, William H. Long, and Isaac Copeland, ran such a pack string, and on March 22, 1875, on their way down from Globe, camped for the night at the base of the hillock. In the morning one of their animals was missing. The better to get a good view across the valley, Copeland scrambled to the top of the little rise. He promptly stumbled across a ledge of silver lead, which was then and there jointly staked by the four as an association claim and given the name Silver King.

Losing interest in ranching and packing, the partners devoted themselves for fifteen months to skimming off the cream of the outcrop. It was easy work, requiring only open-cast quarrying of the ore, a little upgrading by hand, and direct smelting by charcoal in a crude adobe furnace. The silver bullion they ran was then packed out to the Colorado River and taken by ship to San Francisco. It was estimated that the partners divided about one million dollars, a handsome return on a very small investment.[2] By June, 1876, the work was becoming harder and less rewarding as the pit deepened and the ore thinned. As a result, Copeland and Long sold their interests to Regan and Mason. They, in turn, seeing that regular mining and milling, requiring a great deal of capital and expert direction, was necessary, disposed of the location to the Silver King Mining Company of California, promoted and headed by Colonel (an honorary popularly bestowed on mining promoters) James M. Barney, of Yuma, Arizona.

Barney's people sank a hoisting incline on the outcrop ore, built a stamp mill at Pinal, and operated the development to such good effect that the company grossed an additional two million dollars in seven years, paying a substantial series of dividends.[3] Exploratory work revealed that the lode was a pipe or chimney, roughly cylindrical in its eighty-foot cross section, with its axis dipping slightly to the west. The country rock of the basin was a typical purplish porphyry, which had been replaced in the lode with shoots, stringers, and threadlets of quartz and barite, carrying the silver lead with a few traces of copper. It was a real pleasure to mine, since the country prophyry was sound rock which had also sealed off surface water so well that the Silver King was virtually a dry mine. In 1879 a big vertical hoist shaft was sunk a hundred yards west of the outcrop, and some time later a second shaft was sunk a quarter mile beyond, undoubtedly for ventilation and safety and to cut down on underground tramming.[4]

Silver King ore is very handsome, consisting of a gangue of opaque white cryptocrystalline quartz, as attractive as Bristol glass. It includes occasional patches of creamy barite, while the ore particles are either silvery or a brilliant iridescent blue. In-hand samples may serve without further attention as cabinet specimens, while the early "jewelry" ore containing sprays of native silver must have been a joy to behold, in contrast to ordinary gold or silver ores, whose attractiveness is mostly inferential. The best place to prospect for such specimens here or in any mine yard is neither at the headframe dump, which has always been well picked over, nor at the mill tailing heap (should it still exist), but along the shoulders of the haulage road which led from the headframe to the mill. As with any more traditional prospecting a search goes best on the first sunny day following a good rain.

Haulage-road prospecting is rooted in the fact that the ore wagons lurched and jolted in the ruts, shaking off a certain amount of ore from the load. Still other chunks would have been picked off the load by the teamsters to shy at the hindquarters of whichever mule was shirking its share of the work. It is said on good authority that, as a given teamster was passing his own home, he would invariably find fault with the entire team, tossing a continual stream of the highest-grade ore chunks he could find within reach. If then his family came out and retrieved the ore, ultimately selling a wagonload on its own account to the mill, it is evident that the real moral gravamen must have laid with the shiftless mules.

Milling Silver King ore presented some technical and economic problems despite the fact that silver lead in a quartz gangue is theoretically one of the most readily milled of all minerals. The absence of cheap transportation meant that only thirty tons of the highest tenor ore could be stamped, concentrated, and smelted each day, while the middling-and-low-tenor ore had to be impounded pending more favorable conditions. When the railroads arrived in Arizona in the 1880's, this stockpiled ore was run under the stamps and shipped to San Francisco as concentrates to much greater economic advantage. Too, silver-lead sulphides are remarkably given to "floating" in the presence of the slightest trace of grease or oil and those dustlike sulphides which went over the gravity-concentration equipment were considered lost for good. The heavy but relatively worthless barite further complicated matters, but by 1883 the mill at Pinal was recovering 92 per cent of the assay values, a not-so-bad degree of efficiency for the day. Unfortunately, about the same date the lode values thinned out abruptly and quit for good around the seven-hundred-foot level. The Silver King was played out, and although subsequent efforts were made to explore deeper and to scavenge the waste dump and tailing heap, Sullivan's discovery by 1893 was dead.[5]

Now and then the dump of a played-out mine was filed on and worked at a handsome profit by a former employee. It almost always happened that this man had been in charge of sorting ore from waste as it was delivered by the cage or skip at the headframe. His official duty had been to send waste to the dump, reserving ore of economical tenor for the haulage wagon bin. If, however, he had reason to suspect that the mine was playing out and would shortly be abandoned, he might not be quite as conscientious as his employers were paying him to be. The result would be that a sizable tonnage of pay ore would go to the dump, perhaps even to one particular spot on the dump, since the man himself would probably be pushing out the cars of waste. After the mine had closed down, he could then return, file location papers on the dump as a mining prospect, and retrieve quite a bit of very profitable ore for himself.

Even though a silver outcrop showed no nuggets of native silver lead, the experienced prospector had other strings to his bow when it came to investigating it. Both iron and copper are usually found in small percentages in silver lodes, and these trace metals weather down to produce colors which are visible from a distance and are very characteristic. Iron stains deep brown to dead black, as does the manganese which often accompanies it. Every European language has the equivalent of the Cornish expression, "Iron rides a good horse." Rickard speculated that the silver lodes of Laurium, whose country rock composes a sea cliff, may first have been noted by Phoenician sailors, whose attention would have been drawn to the black stain originating from the highest ore horizon.[6] Weathered copper produces a brilliant blue-and-green crust which is unmistakable yet which often leads greenhorns to excessive zeal. It does not take much copper to produce a very theatrical "blossom" of this sort, hence the cautionary proverb, "A copper penny will stain a square mile." Nonetheless, these colors pointed to some degree of mineralization, and were considered signposts well worth investigation.

Good surface color and a substantial drag of silver chlorides in the prospect pan did not necessarily mean that the finder's fortune was made. The West abounds with small lodes which outcrop in such a manner that atmospheric action has naturally milled and concentrated their scanty primary values into a very high tenor but very limited point deposit, or ore pocket. This surficial pocket, perhaps no more than two or three cubic yards of pay ore, perches on a stringer of primary ore of very uncertain worth and limited extent. Such phenomena are common in the Huachuca Mountains of Arizona, where the prospects are said to lie "big end up." Encountering one of these and suspecting that his discovery would not stand blasting, the old-time prospector was inclined to leave it alone, lest he undermine it in more senses than one. It was as good as money in the bank as it lay, while there was a slight possibility that an optimist unfamiliar with the country might come along to buy the location at a price inflated by hopes.

Occasionally such tactics came to a bad end. Thomas A. Rickard told of meeting an old Negro prospector wandering dazedly in the pine woods near Breckenridge, Colorado. When Rickard asked him if he had lost his way, the prospector denied that he ever got lost.

"Where's your prospect?" Rickard asked next.

"There," the old-timer replied. "There's the best mine in the State— ruined by that damned nigger in Denver."

"How's that?" Rickard inquired.

"He put in one shot and blew out all the ore," replied the unfortunate owner of the new *borrasca*.[7]

The bond-and-lease system was evolved around the 1890's to protect buyers from mines allergic to dynamite, as well as from those upgraded in a manner contrary to nature by a bit of judicious salting. Under this arrangement the prospective purchaser posted a bond in the amount of the selling price. He then took possession of the location and financed the necessary development work. He sold all pay ore which development revealed, pocketing the proceeds. At the expiration of the lease period, should the mine continue going strong, the purchaser could exercise his option to buy, with the bond money going to the vendor. If, in the lessee's opinion, the prospect was turning out badly, he returned the location to the vendor complete with development installations, reclaimed his bond, and departed in peace.

Virtually every major mineral district in the West was initially located by experienced prospectors who put in a great deal of time and effort before their strikes proved out.[8] Now and then, however, undeserved luck played its part, and in one case whisky-jug prospecting paid handsomely. Horace A. W. Tabor, of Leadville, Colorado, undistinguished either for intellect or industry, kept a grocery store in the locality to cater to the mineral seekers of that newly discovered district. His geological knowledge, observed the humorist Bill Nye, "had previously been in the great field of family mackerel and the study of rock salt and codfish float,"[9] but in 1878 he grubstaked two prospectors, August Rische and George F. Hook, who wished to seek the silver lead ore of this region. In addition to food Tabor threw in a bottle of whisky as a tonic and appetizer. Rische and Hook had traveled barely a mile from Tabor's store when they decided to consume the bottle before its contents spoiled in the high altitude. The site of their aperitif was Fryer Hill, on which some prospecting work had already been done to uncertain advantage. Deciding after their refreshment that one spot was as good as another, the partners stayed where they were, put in location monuments, named the site Little Pittsburgh, and began to sink a shaft. At twenty-six feet they hit a substantial stratum of silver-lead carbonate ore, the first wagonload of which was sold for over two hundred dollars.[10]

Tabor's luck stayed with him. A short while later a Denver whole-sale house with which he did business asked him to buy them a likely prospect. Tabor laid out $40,000 for the Chrysolite location, a flooded shaft in a nondescript limestone stratum which had been salted with ore from the Little Pittsburgh itself by "Chicken Bill" Lovell, a noto-riously evil liver. Lovell boasted of Tabor's gullibility, the word got back to Denver, and the wholesalers indignantly repudiated the trans-action. Unperturbed, Tabor ordered development to proceed, and a few feet farther down the shaft hit an ore body of higher tenor and greater extent than the Little Pittsburgh itself. Tabor and his backers took out $1.5 million before selling the Chrysolite for as much again to another syndicate.[11]

It is fairly obvious that silver-lead ore which can be struck by more or less random sinking must differ markedly from such a primary lode as the Silver King. As a matter of fact, these Leadville ores were "replacement" deposits, being found in a blanketlike sheet on the sur-face of a blue limestone which was completely unmineralized when it was first laid down at the bottom of a warm and probably shallow sea. As the land rose and buckled, the limestone was uplifted and deeply overlaid by other strata of sediments. At some subsequent period igneous action forced a broad intrusive stratum of green-white por-phyry across the upper surface of this limestone; the porphyry carried silver-lead sulphides, although in very low percentages. As the por-phyry cooled, it lost up to 10 per cent of its volume, causing it to become cracked and fissured extensively. This afforded avenues for the percolation of ground water, which dissolved and carried down the values. These values were chemically precipitated in high concentration on the contact zone, where the porphyry was bedded on the limestone. Within this zone the limestone was dissolved by the acid water and physically replaced by the silver lead to form the blanketlike ore sheet.[12]

During this replacement process, the metal values were transformed from the primary sulphides to silver chloride and lead carbonate respectively, collectively termed *carbonate* ore. Other metals present in trace quantities did not necessarily halt there, but could be carried on and deposited elsewhere. The original discovery of this stratum of carbonate ore was made by two old prospectors, W. H. Stevens and A. B. Wood, who were cutting a water-supply ditch for the placer camp at California Gulch and who had the intelligence to investigate more thoroughly the dark, soft, heavy mineral which their picks were casting up. This mixture of silver chloride and lead carbonate in a limestone gangue happened to be just what the Boston and Colorado Smelter of the Central City District needed badly to flux the refractory pyritic gold ores of the Colorado Front Range. Such flux materials are called smelting ores. They commanded premium prices because of their utility, prices which in fact often exceeded by a considerable sum the bare assay value of the included metals.

A much more complex replacement body of carbonate smelting ores was found at Tombstone, Arizona, by the prospector Edward S. Schieffelin after a particularly prolonged search. Fortunately for history, Shieffelin left an extended account of his adventures, including an autobiographical sketch which indicated that he was born in October, 1847, probably the third of the six children of Mr. and Mrs. Clinton Schieffelin, of Tioga County, Pennsylvania. In 1852 the elder Schieffelin moved to California, doing well enough there to send for his family four years later. In February, 1857, the reunited family settled on a farm near Jackson City, Oregon. The boy Edward began to contract a chronic case of prospecting fever, perhaps from exposure to the tales of the forty-niners of California, but more likely by news of his maternal uncle, Joseph Walker, a gold rusher and an incorrigible boomer. At the age of ten Ed was panning yellow mica from a nearby stream, and two years later he ran away from home in an attempt to get in on the Salmon River placer excitement. Sixty miles from home his enthusiasm wore off, and he was content to return to the hogs and hay until he reached the age of seventeen.

In 1864, Ed was a powerful young man nearly six feet tall, weighing 175 pounds, and with the beginnings of what would later be a luxuriant brown beard. He left home again and, being a capable teamster, fell into the prospectors' habit of working at his trade long enough to build up a stake before quitting to join the latest mineral excitement. He moved about, to Nevada, the Owens Valley of California, the Salt Lake rush of 1871, and Boise, Idaho. He circled back to Eureka, Nevada, where, he said, "for eighteen months I worked hard, and tried several different things, but although I was economical, neither smoked nor drank, nor gambled, nor spent money in unnecessary ways. . . . I found I was no better off than I was prospecting, and not half so well satisfied." Since his "several different things" included teamstering, cutting cordwood, and doing mine work, his attitude is understandable. He also became convinced that following rushes was far less promising than striking out on his own into new territory.

In November, 1875, suffering from a fever and cough, he returned to his family home with only $2.50 in his pockets to show for six years of wandering. Three weeks of rest and home cooking cured his illness but not his wanderlust. Ed borrowed $100 from his father and went to the borax camp at Ivanpah, California, where he worked for fourteen months, increasing his stake sufficiently to buy a prospecting outfit, food, a pack mule, and a riding mule. In January, 1877, he left San Bernardino for Arizona and the great adventure of his life.

Prospecting outfits varied according to the expertise and the tastes of the individual, as well as the availability of transportation. Such equipment as the essential pick, the engineer's "round-point" shovel, and the eighteen-inch prospector's pan were available at every store in the West. Other items could be purchased at an assay house or could be picked off the town dump and suitably modified by an obliging blacksmith. The field mortar, created by sawing off the bottom third of an iron mercury flask, was needed; the pestle was any large bolt or a foot length of "Black Diamond" drill stock. The horn spoon, needed for delicate washing of panned concentrates, was also made by sawing a small iron skillet in half. Acid, charcoal block, blowpipe, and mercury were used by those who ran their own field assays, the liquids being carried in the old-fashioned glass beer bottles with ceramic stoppers held in place by a wire toggle harness. A reel of Bickford fuse, some cans of blasting powder, and short drill steel were highly recommended. Since Schieffelin was not much of a field assayer by his own inferential admission, and since he planned to travel light and fast, his equipment was minimal, supplemented only by his needle gun, a primitive bolt-action rifle necessary for protection and game hunting.

He rode into Arizona in early spring with some thought of prospecting the Grand Canyon but quickly abandoned the idea and drifted southeast through the central mountain belt, traveling in company with a party of Hualapai army Indian scouts for protection against the Apaches. About April 1 he arrived at the newly established Camp Huachuca in extreme southeastern Arizona Territory and soon gained employment as a guard for the small crew doing the annual assessment work on the Brunckow Mine, a minor silver lode near the military post. The necessity of a guard was shown by the fact that no less than seventeen men, including Frederick Brunckow himself, had been killed, mostly by Apaches, at or near the site.[13] The site of Schieffelin's guard post is uncertain. The lode itself is on the flat, affording poor observation, and there is a quite high, conical hill a mile away, commanding a magnificent view, but it would be very difficult for a lookout on the hilltop to attract the attention of men working in the Brunckow shaft.

The locale of the Brunckow Mine lay a trifle over a mile from the meandering course of the San Pedro River in the midst of an extensive, gently rising plain. Although seemingly quite open, the plain is marked by gentle, brush-filled undulations affording good cover for a war band to work in too closely for comfort. Whatever his position, Schieffelin swept this plain with his field glasses, presently noting parallel "dragon's teeth" outcrop ridges running from the Brunckow location toward a line of low, rolling hills eight or nine miles northeast from his stand. It occurred to him that these linear croppings might have been the conduits of silver mineralization. On this assumption he made several foot scouts toward the hills, during which he found attractive float of some sort, either rusty quartz or heavy black mineral; either was possible, but Schieffelin neglected to be specific.

One of the assessment crew, William Griffith, learned of Ed's interest in the hills and offered to grubstake a survey of the formation if Schieffelin would make a joint location with him on anything he might find. This would take the form of Schieffelin staking two claims and placing the end-line center marker between the two at the point of discovery—a fair and equitable arrangement which suited Schieffelin well. On the strength of this he went over the deeply gullied plateau during the summer of 1877, finding more float, metallic nuggets, and, at last, an ore stringer on which he staked the Tombstone and the Graveyard locations. This was a sarcastic reflection upon the Camp Huachuca garrison's prediction that he would find nothing in this Apache-terrorized country but his grave. Unsure of the composition or tenor of his samples, he rode in August to Tucson, a squalid collection of adobe hovels then enjoying, or failing to enjoy, its deserved reputation as a paradise of devils. He recorded the locations in his and Griffith's names, but found that the town had no assay house. The locals dismissed his specimens as worthless. Unable to find help or encouragement, Schieffelin returned to continue his prospecting through the autumn, despite an Apache scare of even greater magnitude than usual.

In retrospect it can be seen that most of Schieffelin's trouble lay in his assumption that he was prospecting a primary lode similar to the Brunckow, whereas in actuality he had come on an extremely complex replacement deposit—nothing was where it ought to be, by Schieffelin's lights. The long parallel ridges which had first drawn his attention were porphyry dikes which had invaded and traversed the Tombstone hills, a heavily eroded plateau of heaved and faulted strata of granite, limestone, quartzite, shale, and serpentine. The silver-lead values had migrated from the porphyry and had been deposited in secondary concentration in the tops of the arched limestone folds (anticlines). Although heavily mineralized, these folds were hidden beneath a jumble of scree. Schieffelin was walking over silver lead almost everywhere he went, but its surficial signs were indirect and misleading; for that matter, the Tombstone-Graveyard location was only a thin ore stringer which was never profitable to work. On the other hand, Schieffelin was stubbornly convinced that his indications did not lie, and he continued to press a search which less determined men might by then have abandoned.

About October 1, with food and funds exhausted, Schieffelin decided to enlist his older brother, Albert E. Schieffelin, whom he thought to be working at the Silver King Mine. Ed rode north to the Pioneer-Globe District, arriving at Pinal so broke that he spent there his last twenty-five cents for tobacco.[14] He found that Al was not there, but was reported to be working at the Signal Gold Mine in the Kingman District of northwestern Arizona Territory. This was a ride of several hundred miles which Ed preferred not to undertake on an empty stomach with winter coming on. As a temporary measure he took a job as a windlass man at the Champion, a small silver mine twenty-five miles north of Globe.[15] His working conditions were poor, as befitted an evident ten-day miner who only wished to raise a stake: he was put on the night shift, had no shelter from the cold wind, and was paid but three dollars a day.

At the end of the traditional two weeks Ed drew his time and headed northwest to the Signal Mine, where he found Al, the steady member of the family, who got him work as a shaft-sinking mucker at four dollars a day—the hardest of work, but better paid on that account. Ed at once began trying to convert his brother. The conversation paid unexpected dividends, for it attracted the attention of Richard Gird, an assayer who himself had just reported to work at the Signal. Gird asked for the ore specimens, assayed them, and found them to be virtually pure silver lead. A sample from the Tombstone went eight hundred dollars to the ton, while another from the Graveyard ran as high as two thousand dollars. Gird became enthusiastic, talked Al into partnership, and about February 1 the three left the Signal with a wagonload of supplies and Gird's assaying equipment.

The partners arrived at the Brunckow on February 26, making their base camp in an old adobe hut, around which seventeen dead men were reported to have been buried. A few days later they woke one morning to find that their draft animals had been stolen by Indians. Since this was the usual preliminary to an "Apache haircut" in the territory, the partners understandably became somewhat nervous. Gird remained indoors to construct an assay furnace, while the brothers went out to hunt and scout for Indian sign. Soon, however, Ed's feet drew him toward the Tombstone hills. In early March he came across a broad fan of brown iron stain which originated under a covering of loose scree, approximately in the center of the plateau.[16] Grubbing into the confused mass, Schieffelin found a black ore ledge of argentite or cerargyrite that was seven inches wide, could be traced for fifty feet, and was so metallic that it could be "printed" or embossed with the head of his pick. Gird took the samples in hand and soon reported that they could go as high as twenty thousand dollars, of which twelve hundred dollars was gold. This location was promptly staked as the Lucky Cuss.

The prolonged activity of the partners attracted the attention of other prospectors who had drifted in, scenting action. Rather than fight competition, Gird co-opted it, offering free assay facilities to all comers in return for a half interest in any good locations they might make. Possibly Ed Schieffelin shared with them his hard-gained knowledge of the local signs of mineralization, but with the help of his experience he found in April black iron-stained croppings at the northeastern rim of the plateau. They too assayed well, and the Tough Nut location was added to the partners' accumulations. One of the newcomers, Hank Williams, discovered the head of the Grand Central anticline on the east rim, but was at first disinclined to honor the share agreement. Gird remonstrated with him to such effect that Williams moved his end-line monument three hundred feet south, permitting the partners to run a location north from this boundary. It was named the Contention in memory of the dispute, and the Contention–Grand Central was to become the greatest of the Tombstone ore bodies, although geologically jinxed—later development revealed that the main ore shoot had been pinched out five times in succession by earth faults.[17] To raise development money, the Contention was at once optioned for ten thousand dollars for sixty days to two other prospectors, White and Parsons.

Now that the Tombstone plateau was reasonably well explored, there began the wearisome business of development on the one hand and of negotiating acceptable offers from mining syndicates on the other. The partnership's man of business, Gird, here came into his own. His major coup was to talk Territorial Governor A. P. K. Safford into the combination in return for financing a ten-stamp mill of modest capacity. It was cast at the famous Fulton Iron Works of San Francisco, freighted to Tombstone, and erected at the San Pedro River, which provided enough water for the concentration tables and engine boiler. The mill went on line June 1, 1879, and the partners them-selves hauled the first eight bullion bars to Tucson on June 14.[18] Beyond the satisfaction they had in pouring their own silver, the income from the mill helped retire the partnership's debts and put them in a much better bargaining position with the numerous, but mostly unsatisfactory, tenders for this or that portion of their sheaf of locations.

Unable to conclude a sale locally, Gird advised that the Contention location and the mill be incorporated as the Tombstone Mill and Mining Company. Al Schieffelin and Governor Safford traveled East to sell shares in the venture, while Gird managed the mill and Ed vanished into the hills to hunt and ramble, having found development and promotion too boring to endure. He returned from his vacation to find that in February, 1880, a Philadelphia syndicate had come up with an acceptable offer. The partners split $600,000, with Ed and Al taking cash, but Gird foresightedly accepting his share in stock. Their other locations were disposed of to advantage, and thanks to Gird's sound guidance the Schieffelins became moderately wealthy men. The prospectors bowed out; it was now time for the promoters, engineers, and millmen to move in, making Tombstone the bonanza camp it would remain for ten years.[19]

Development on the plateau soon revealed a novel and almost unprecedented geological situation in the ore bodies. It is most unusual for a good fluxing ore and a refractory ore that requires fluxing to be found in the same deposit or even the same district, but Tombstone proved an exception. The gold values in the primary porphyritic dikes were much richer than might be expected, and yet there were no appreciable percentages of zinc, copper, or arsenic in any of the ores. The absence of these latter values strongly hints that there may have been two distinctly different periods of mineralization: the silver lead, which invariably appears first, followed perhaps much later by the gold-iron-manganese phase. Atmospheric action had taken the silver lead out of the dikes and deposited it as a replacement carbonate ore on the limestone anticlines, but it had left the gold values in place and reduced the iron pyrites, which carried them, to hematite. This last was so pronounced that the miners in the gold horizons came off shift powdered with iron oxides until they looked like red men.[20]

In consequence of all this, Tombstone was remarkable in that one mine might be hoisting lime carbonate silver-lead ore while its immediate neighbor would be working a gold quartz porphyry. Worse for the gold producers, the weathering had coated the free gold particles with iron silicates and black manganese film, rendering them resistant to the (then) standard milling practice of recovery by amalgamation with mercury. Such refractory gold ores could be readily smelted when fluxed with carbonate or manganese ores. Since the camp providentially was producing these fluxing ores, things came out more or less even in the long run.

At the hard-boiled mill town of Charleston, Arizona, smelters soon blossomed beside the concentration mills along the San Pedro River. Ores freighted down the long gradient were crushed, stamped, and concentrated by tables or vanners. The coke and such other necessary materials as pyrites and quartz sand were added to carefully adjusted batches of smelting ore and refractory gold concentrates and the fires lit. The easily reduced silver lead descended in droplets through the charge, collecting, fluxing, and carrying down the gold, while the quartz and iron formed a slag in which collected all other impurities that did not depart as smelter gas. Extra carbonate concentrates were bagged and shipped out by railroad to other smelters as far away as Utah and Texas. Even though the tenor of all the ore horizons began falling off rapidly as the very high water table was hit, the Tombstone mines are estimated to have hoisted in the aggregate about six million dollars a year during the decade of their greatest activity.[21]

Photographs taken of the camp at its flush period fail to convey the air of intense activity and excitement which permeated Tombstone or, for that matter, any booming mineral development. They cannot depict the woodsmoke drifting from a hundred engine boilers, the constant clanging of the hoist signal bells, the clamor of the smithies, the erratic whirling of the headframe sheaves as the hoist cages rose and dropped, the twice-daily rumbling of the ground as the drift rounds were fired, and, most characteristic of all, the chant of the Charleston mill stamps produced by fifteen hundred pounds of cast-iron stamp shoe dropping every half second, night and day, on the ore. In addition to these recurring sights and sounds, there would be the braying of ore-wagon mules, the clatter of waste being spilled over the dumps, and the melodious whooping of mill and mine whistles. To modern ears, however, there would be a strange omission in the middle registers of sound—steam engines ran in ghostly silence, and it would be fifty more years before the most characteristic noise of a big mining operation became the harsh, unending blat of diesels and accompanying groan of overloaded power transmissions.

Ed Schieffelin did not stay around to participate in the technical and social uproar which his vision had brought to the windswept plateau and to the camp on Goose Flats at its base. He and Al (who had been increasingly ill) went their ways, Ed to return to Jackson City, where he bought his family a home in Los Angeles and moved them down into the sunshine. He himself went to Alaska following persistent rumors of placer gold on the Yukon River, but he neither made a strike nor had any success in running a riverboat, the *New Rocket*, which he had bought. In 1883 he returned to the States and at La Junta, Colorado, married Mrs. Mary E. Brown, presumably a widow. The next year his father, Clinton Schieffelin, accidentally shot himself to death. Ed and his wife moved from their home in Alameda down to Los Angeles to keep his mother and the rest of the children company. In 1885, Al came home—to die of tuberculosis. Perhaps feeling a little lost, Ed continued to go occasionally on prospecting trips. He still appeared healthy and strong, but during one such trip to Oregon in 1897, he was found dead of a heart attack at his little cabin. He had packed a great deal of living into exactly fifty years, pursuing the gold bug with courage, honesty, and dogged determination. As much as any man, Ed Schieffelin earned the traditional frontier epitaph: "He done his damndest; angels could do no more."

CHAPTER V

Rope and Bucket

THE transportation of ore and waste from the face of the heading to the reduction works or dump has always been of concern to mining men. Until the development of blasting, however, the attention devoted to this problem was slight in comparison to that given other matters; the quantity of rock that could be daily broken out by cold mining or firesetting was relatively trivial, and the cost of transporting it negligible. For that matter, the miner working in really hard rock could usually remove in a sack or basket all that he had broken out during his shift. Whenever this task was delegated to porters as such, it was usually a consequence of large-scale working in more or less friable ground. This division of labor enabled the management to economize on the physical strength expended in mining and crushing or on the skill of those working in the concentrator and smelter. The Greek geographer Agatharchides noted of the New Kingdom gold mines in Egyptian Punt that the broken gold-quartz ore was carried out of the drifts by boys who had not yet come to maturity,[1] a practice which still persisted in the collieries of early Victorian England.

Such porterage required very little in the way of special equipment or arrangements. Until the late Middle Ages vertical ascent was made by "chicken ladders," deeply notched logs set at a decided angle to the sides of the shaft, their ends squared and secured in mortises, or "hitches," chipped out of the rock. Though crude in appearance, they were stronger than conventional ladders and easier on the arches of poorly shod feet, while the slant enabled the porter to lean forward so as to balance the load on the small of his back. A tumpline, or forehead band running under the bottom of the container, anchored the load and freed the porter's hands. Broken rock was carried in a wicker, hide, or fiber basket, while sump water was transported in a waterskin. Porterage in this fashion was employed in the silver mines of

Periclean Athens[2] and persisted as late as 1860 in the more backward developments on the American mineral frontier.

An unedifying story is told of a visitor to one of these American frontier mines where bailing was still being conducted by porters using waterskins.[3] The visitor wondered aloud how these skins were obtained, as they obviously were, in one uncut piece. He was placidly informed that the pig in question was systematically starved for some time so that its skin became quite loose. One end of a stout line was fastened to the pig's tail and the other end securely tied to a tree. An ear of corn was then waved in front of the pig's nose, so inducing the animal to slip out of his skin and walk away from it. The visitor appeared to entertain doubts about this process but ultimately accepted it as true.[4]

The widest use of such primitive methods occurred in the deep silver lodes of Spanish colonial Mexico, where the *tenateros*, or mine porters, were famed for their ability to carry as much as 350 pounds deadweight upward for many hundreds of feet in their *tenates*, or fiber baskets. The Indian porter leaned forward against his tumpline, steadied himself with his right hand, and in his left hand carried a wooden stick which held his candle in the end. When bailing, his pigskin waterbag was supported in exactly the same way. It is probable that a Neolithic flint miner would have found no change, apart from the candle, in the equipment and dress of his Mexican counterpart forty-five hundred years subsequent to his own era.

In classical times the lowering of such materials as timbers might be done by two-man carrying teams, although where shafts existed, as at Laurium (Attica) swaying them down with fiber or rawhide lines unquestionably suggested itself. The ancient peoples used a great deal of rope in seafaring and construction work, but the endurance of the material (papyrus, flax, or hide) was poor. Their hoisting gear could be highly developed; a relief of Imperial Roman times clearly shows compound pulleys being used in erecting cranes, the power supplied by treadmills. Their use of cranks and windlasses is uncertain. Most historians of technology assert that cranks were not invented until about the ninth century A.D.,[5] but certain evidence points to at least limited use at a much earlier date.[6] Archimedian screw pumps, chain pumps, and the *noria* ("water-lifting wheel") were used in the classical world,[7] and so it is a fair assumption that a certain amount of semimechanized haulage was within the capabilities of the ancient miners. And if materials could be lowered, ore, water, and waste could be raised.

Whether this possibility was common practice might be discounted by various negative considerations. The habit of sending the most brutish or dangerous slaves to the mines would certainly cause a supervisor to think twice before installing equipment whose potentiality as a murder weapon was considerable. Even proper blocks and sheaves would have subjected the weak hoist ropes to undesirably prolonged sharp bends and dangerous frictional wear, much shortening their life and increasing the danger to those working under the bucket. Indeed, the equipment illustrated as late as the publication of *De re metallica* has nothing that can be interpreted as a block or sheave. Hoisting and running ropes are wrapped about drums which in some cases are grooved longitudinally to afford a better frictional purchase. The question must remain unresolved whether these hoist devices were unknown, despite their wide use at sea, or whether they were avoided because of the much greater dead loads encountered in mining work.

The technological upsurge in the Western world that followed the fall of Rome is indicated by the wide display in *De re metallica* of materials and methods which were unknown or unusual in the classical world. Wooden-stave casks of Teutonic origin, horse harness which is based on the singletree and the horse collar, extensive use of wrought iron, valved lift pumps, and even centrifugal ventilation air pumps are all shown clearly. These were not invented overnight, and the list may be extended to include such developments as spring steel, the vertical windmill, and a wide variety of industrial processes ranging from movable-type printing to assaying. By Bauer's time in the mid-sixteenth century these devices were being used routinely in a society in which the miner was a free and prideful technician, aware of his social usefulness and of the need to make brains an economical and practical substitute for brute strength. If a Saxon *bergemeister* could think of a better way to do the job, his financial backers would not say him nay, nor would his crew instantly down tools in protest against an assault on craft traditions.

Although the equipment shown in *De re metallica* is still a bit rough around the edges, the general trends are surprisingly forward-looking. Perfectly good double-crank windlasses are on display, their standards

Primitive bailing and hoisting buckets. *L*, various sizes of hide bailing buckets. *C*, wooden-stave buckets, prone to slip around on the hook because of the lack of a loop in the center of the bail. *K*, a wicker basket whose flattened side appears woven to fit a porter's back. From Georgius Agricola [Georg Bauer], *De re metallica*.

even having slots out of which the crank and drum can be lifted and set aside on the platform collar. This is almost uncanny, for blasting was not yet developed, and yet it is necessary to remove the windlass in this manner before a shot is fired so that it and the rope will not be cut up by flying muck.[8] The same woodcut shows another windlass set at the collar of a winze (internal shaft), while two trammers are each pushing four-wheel ore cars down drifts. These cars run on planks for greater ease, the car being kept railed by a guide peg running down into the space between the planks. The sinking buckets are made of wooden staves, keg-fashion,[9] and in another illustration two of them are set in counterbalance so that as one descends the other rises.[10] The presence of runged ladders shows that porterage has been almost totally abandoned in favor of hoisting, and a variety of buckets are illustrated: animal-hide, which is best for sump bailing,[11] iron-strap reinforced wooden-stave buckets for ore hoisting, and also a wicker basket with one flattened side which looks suspiciously as though it had been woven to fit on a porter's back.[12]

In an age which was rapidly developing wind, water, and more effective animal power, Bauer uses every chance to display the application of these to mining, his favorite prime mover being the efficient overshot water wheel. These drive stamp mills, ventilation fans, and a wide variety of lift and suction pumps, but pride of place is assigned to a big, businesslike shaft hoist which is really ingenious (whether it was ever installed or was merely suggested). It is powered by two overshot water wheels, constructed as one unit on a common axle but with their buckets set in opposing inclinations. The axle is prolonged into the hoist drum, which has a grooved "wildcat" to engage securely the links of the chain hoist rope. One end of the chain is raising a large hide bailing bucket, while a hook on the free end of the hoist rope is displayed, clearly indicating that another bucket may be attached thereto to gain counterbalance in the most approved fashion. The operating engineer sits in a cupola overlooking the shaft, manipulating watergate levers; one opens the flume to drive the wheels in one direction, and the other opens to reverse the rotation of the wheels and drum. About all that seems to be missing to the modern eye is a hoist-shanty bell-code placard and a quid of tobacco in the engineer's cheek.[13]

De re metallica served as an instruction manual and set of construc-

A water-powered hoist. *A*, reservoir; *B*, race; *C*, *D*, levers; *E*, *F*, troughs under the water gates; *G*, *H*, double rows of buckets; *I*, axle; *K*, larger drum; *L*, drawing chain; *M*, bag; *N*, hanging cage; *O*, man who directs the machine; *P*, *Q*, men emptying bags. From Georgius Agricola [Georg Bauer], *De re metallica*.

A sectional view of Saxon mine workings. *A*, shaft; *B*, *C*, drift; *D*, another shaft; *E*, tunnel; *F*, mouth of tunnel. From Georgius Agricola [Georg Bauer], *De re metallica*.

tion designs for the Saxons who emigrated to foreign mining camps during the seventeenth century. Many German technicians went to Spain and to Latin America, but the comparative scarcity of wood and waterhead in these semiarid regions (plus a real shortage of iron in the New World) forced extensive revision of the Saxon mining canon. This revision was done very intelligently, however, and mostly by the Spaniards themselves; thus their *sistema del rato* ("system of the moment," that is, "empirical system") persisted in Latin America for over three centuries, during which time it hoisted an estimated three-fifths of the world's present silver supply.[14]

Those Saxons who emigrated to work the copper lodes of Cornwall at the behest of the Tudor monarchs found a climate and resources which approximated those of their homeland. Their methods were fastened upon and considerably improved by the Cornish, themselves enthusiastic placer men and miners of tin. The Cousins modified Saxon windlass hoisting in some small but significant details. Being great seafarers, they introduced hemp ropes which were tarred heavily to minimize rot. The bucket, or "kibble," was fashioned of riveted sheet iron, and its profile was changed from a truncated cone to ovoid, the better to prevent it from hanging up on rocky projections in the shaft. The iron bail was given a center loop to prevent the hook, or eye splice, from slipping around, and the bottom was provided with a down-haul ring and short rope both to facilitate pulling the kibble over to the shaft platform and as a point of attachment to fasten long timbers for lowering. This Cornish bucket cost rather more than the wooden-stave bucket, but it was watertight, more durable, and appreciably less aggravating to use.

A peculiarly Cornish development was the vanning shovel, a remarkable advance beyond the circular or elliptical wooden washing bowl of Saxony.[15] Frequent estimation of mineral percentages in ore or mill pulps is essential to prospecting, drifting, and, above all, the economic control of concentration processes. Although this is most accurately done by assay, it may be satisfactorily approximated on the spot much more cheaply and swiftly by taking a sample of pulverized ore and separating it into its components by gentle agitation in a water medium. The bowl was good enough for the Saxons, Spaniards, and Americans, who dealt mostly with silver lead or gold of high specific gravity, but the bowl could separate only head values from tails, while discarding the tails in the process. It was not sufficiently delicate to grade the less massive copper or tin slimes or fines. For this the Cornish developed a flat, round-point shovel of thin sheet iron which in expert hands separated the milled ore into a number of distinct bands. The spectrumlike width of these bands displayed the approximate presence and proportions of heads, tails, and middles (ore of low tenor but still appreciable mineral content which might profitably be subjected to recycling in the concentration plant) in a rough but rapid and reasonably accurate manner.

It was probably also in Cornwall that the headframe, sometimes called the gallows frame, was devised, perhaps originating in the "whip and derry" or simple block lashed to a tree limb extending above the shaft compartment. A hoist rope was run through the block, its free end was attached to the bucket, and its bitter end was secured to a horse's harness. The horse walked away from the shaft to hoist the bucket and was cautiously backed to lower it. In the absence of a suitable tree, the block could be secured to the top of a heavily braced wooden tripod, which gradually evolved into the right triangle rising ten or more feet above the surface of the shaft platform. Nothing greatly suggesting a headframe is shown in *De re metallica*, but its advantages were such that, once invented and properly equipped, it caught on everywhere and remains to this day the most characteristic and visible item of industrial architecture in a mining region.

The attractions of this hoisting tower are many, beginning with the fact that, unlike a windlass, it leaves the shaft collar clear of impediments and permits the toplanders to lower long timbers or pipes and to swing and dump the bucket in any direction desired—ore goes one way, waste another, and water usually a third. It also allows the bucket to be raised much higher than a windlass drum will permit, greatly easing the task of pulling the bucket over to the platform by a horizontal movement in the manner of a pendulum. Likewise it permits the application of draft animals or machinery to the task of hoisting by permitting them sufficient room to work at a comfortable distance from the shaft collar. Not only is clearance for these activities necessary in itself, but the distance diminishes the vibrations which otherwise can dangerously loosen the ground of the shaft wall. Lastly, it is wise (and today in most places it is required by law) that the hoist machinery be well isolated from the shaft so that in case of fire the machinery can still be worked for some time before the engineer is driven from his controls by flame and smoke or the shaft sides begin to cave and collapse. The gain may be measurable in only a few minutes, but in mine fires time is everything.

The headframe greatly increased the capacity and flexibility of hoisting methods in a manner perhaps best illustrated by the huge and complex La Valenciana Silver Mine of Mexico. This enormous and deep-lying lode was discovered by Antonio Obregón in the supposedly mined-out district of Guanajuato toward the end of the eighteenth century. The fractured country shale held so much ground water that extraordinary measures were necessary to make the deposit workable. It was decided to bail the water by sinking a large master drainage shaft, the San José *tiro general*, to the then fantastic depth of fourteen hundred feet. The thirty-eight-foot diameter octagonal shaft was meticulously hewn through the surface andesite by stonecutters who were at great pains to maintain its dimensions, plumb, and surface dressing to very close tolerances. Over the shaft was then erected a great headframe equipped with eight sheaves, each leading a hoist rope to as many different *malacates* (horse whims), each of which had sixteen horses harnessed to its four sweeps. Because counterbalancing of the eight huge hide bailing buckets was not feasible, each *malacate* had to be locked at the end of a lift, the horses turned about and hitched in reverse, and then the *malacate* unlocked to lower the bucket. This enormous and costly work was effective, persisting in use until the political tumults of 1821 forced a half century of abandonment of La Valenciana. Upon the mine's revival in the Porfirio Díaz regime, the shaft was cleared and again put into use with steam hoists lifting self-draining buckets.[16]

In the Cornish mines the problem of water was solved in the eighteenth century by the invention of steam pumping, but the raising of substantial tonnages of ore and waste from depth could not be so readily achieved. As long as the workings remained shallow, porterage with handbarrows was employed, as was "shammeling," or casting up by shovels of the ore from level to level by relays of men. As the workings went below one hundred feet, the cost and the amount of dust

Fall from a bucket. Why Cornishmen did not like to hang from a rope. From Louis Simonin, *La vie souterrane*.

became prohibitive, whereupon hoisting from depth by rope and bucket was introduced. By 1800 it had become a practice to sink a hoist shaft toward which one or more levels of heading would be advanced through the stopes of the long, deep, but narrow lodes. Once the heading had holed into the shaft, the bucket was lowered to its level, pulled sideways into the level station, or "plat," and filled with the ore which had been carried or trammed on the level to that point. When the full width of the lode had been stoped, the miners would install stulls for bracing, these being heavy timbers set at right angles to the footwall and hanging wall—so close to vertical was the dip of the Cornish lodes that the stulls were so near the horizontal as to make no difference. Set as low as possible in the drift, they afforded a convenient foundation for plank flooring. Once this flooring was in place the miners would sink eight feet down and begin drifting on the next lower "floor," raising the ore up to the original heading for loading and tramming to the hoist shaft. The process would go on until too many levels were interposed between the working level and the tramming level, whereupon the hoist shaft would be sunk deeper, the active working level driven into it, and another plat installed.

The Cornish very seldom pumped from the hoist shaft, preferring to keep the two activities well separated, but it was evident that their pump engines had power to spare for hoisting if it could somehow be transmitted to the hoist headframe. This was done by means of long horizontal "flat rods" supported by a series of pivoted stanchions that rocked back and forth as the rod moved. The flat rod terminated in a crank and appropriate gearing at the winding machinery (hoist). This system made hoisting slow, cranky, subject to all manner of unexpected interruptions, and completely unsafe for human transportation. The Cousins did not conceal their aversion to "hanging from a rope" even when the only alternative at the end of an exhausting shift[17] was to climb painfully to grass on perhaps hundreds of feet of ladders. That their attitude was well founded is seen by reference to Louis Simonin's classic *La vie souterrane*, published in 1867. A whole series of illustrations indicates what was happening in the Belgian coal mines, where men rode the buckets. If the ropes or chains did not break, the bucket might overturn, while the presence of flimsy shields set on the rope a few feet above the bucket rim indicates a rather futile effort to divert falling rock from the miners' heads.[18]

The heart of the problem before 1860 was the too-familiar technical fact that increasing loads had greatly surpassed the unknown or uncertain strength of available materials. Hemp or manila rope could not be spun sufficiently strong to support several hundred feet of its own weight together with that of a loaded bucket of any size unless the rope

Riding the bucket. This is a scene in a Belgian coal mine about 1860. Note the bricked shaft walls, miners' safety lamps and flanged helmets, the weak protective shield on the chain clevis, and the young tool nipper with his hand on the bell rope. From Louis Simonin, *La vie souterrane*.

was so large in diameter that it deteriorated swiftly and dangerously when wound on drums or passed over sheaves. The constant unequal bending of the strands soon broke them, while damp rot burrowed insidiously within. Their great cost inhibited frequent routine replacement.[19] Chain was not perceptibly better. Though theoretically much stronger than hemp for the same cross section, the links could seldom run smoothly or lie properly on the hoist drum. When kinked under load, one link exerted leverage that could break the weld of the adjacent link. A few ingenious souls attempted to make rope of parallel drawn iron wires frapped into a bundle by a continuous fiber wrapping, but these wires kinked, broke, and rusted out even faster than the materials they replaced.

At about this time the American mineral frontier opened. Its techniques were an amalgam of Spanish and Cornish methods adapted to meet the scarcity of labor and materials characteristic of a frontier. Windlasses were employed for development shaft sinking to a depth of about fifty feet. Thereafter a crude headframe and a horse whim would normally be installed to raise ore and lower materials. The miners themselves still preferred to descend and leave by ladders, and they cannot be blamed, since even a short accidental drop while riding the bucket could end in a broken back and permanent paralysis. It was yet still taken for granted as well that the only way to transport ore and waste was to have someone lift it by hand from where it had been picked down or shot down, put it into a container, and then hoist that or another container to the surface, where it would be emptied by turning it ninety degrees to spill out the contents. In other words, some man personally lifted every ounce of rock that came to the light of day. In bonanza ore that was almost a pleasure, but by 1870 bonanzas were becoming relatively scarce.

The headframe, hoist, and bucket system was made really practicable by the invention of true wire rope by the German-born John Augustus Roebling (who later designed and began the Brooklyn Bridge). His interest in suspension-bridge construction led him to investigate the behavior of the metal suspenders, previously constructed of chain or eyebars. By 1839 he was experimenting with cold-drawn iron wire and discovered that the only real problem in spinning wire into a true rope was to prevent the individual wires from being twisted

A horse whim. The horse is hoisting two buckets in counterbalance in two shafts. Conditions indicate the abuse to which hemp rope was subjected and thus the reasons for the frequency of accidents. From Louis Simonin, *La vie souterrane*.

torsionally during the manufacturing process. Once this problem was solved, other problems, such as internal corrosion and kink breakage, proved so slight as to need no attention beyond cheap and evident measures, such as galvanization or oil coating. Roebling's wire rope was much smaller in diameter than hemp of comparable strength, was much longer lived under unfavorable conditions (hence over all cheaper), was relatively more flexible, and required no special treatment in the way of splicing, maintenance, or auxiliary fixtures.[20] By 1870 wire rope was being used in the more advanced western mines, where its capabilities inspired high-speed steam hoisting from depths theretofore considered technically or economically unsuitable to work. That being the case, it also set the engineers to thinking about major revisions in the age-old pattern of ore and waste movement.

Peculiarly enough, wire rope in round cross section was almost at once replaced with the braided wire belt for very deep hoisting in western mines. By one account the first such belt was manufactured in San

Francisco and installed in 1863 at the Sierra Nevada Mine on the Comstock Lode.[21] It had been found that wire rope was somewhat awkward to handle with the hoists then in use, owing to the low leverage exerted on it by the hoist drum when most of the length had been payed out. The wire belt was invented (or perhaps adapted from the Cornish flat hemp belt) to minimize this and other problems. Additionally, it was tapered toward the lower end to reduce somewhat the deadweight, and it was wound on wide, thin reels which prevented the belt from uncontrolled crossing as wire rope was prone to do, causing dangerous strains as subsequent turns under load were wound over the "cross-over" section. These reels were so constructed as to give a fine mechanical advantage on starting, with the result that in mines whose shafts extended uninterrupted below one thousand feet the belts for long reigned supreme.[22]

The flat belts caught on rapidly in the bigger mines, where they were greatly favored. Hoist men learned to gauge their degree of fray by holding a handful of cotton waste against the surface as they ran. The amount of waste caught and carried away was a good index to the need for maintenance. Such belts could be unrolled in the mine yard, broken strands removed, and the belts rebraided nearly as good as new. And, at last, they could be cut into sections to be installed as highway guide rails, some doing service as such to this day near Bisbee, Arizona.

The application of the highly developed horizontal-cylinder mill engine required attention only to comparatively minor details. Reduction gears permitted the engine to run at desirably high speeds, while increasing considerably the power available to turn the rope drum or flat reel. Clutches and brakes were essential, and most often were designed on the contracting-band system, in which friction shoes were made of wood: in the Cripple Creek labor troubles a whole cageful of nonunion miners were killed when someone deliberately greased the brake shoes.[23] A governor was not necessary, but a flywheel or two were usually put somewhere in the drive train to provide the initial starting impetus. An indispensable accessory was the "indicator," a clock handlike pointer driven off the hoist by a high-reduction bevel gear. It circled about slowly in response to the number of revolutions the hoist drum had made; hence it could display the approximate footage of rope payed out, indicating the depth and therefore the location of the bucket or cage. The circular clocklike "target" over which the pointer revolved was calibrated by taking the cage down to the level in question, easing it up and down by bell signals until it was in position, and then marking the target appropriately with paint at the tip of the indicator hand. For complete precision a band of paint or a circle of string was put around the rope at a point in line with the operator's eye and some fixed feature in the hoist shanty. An experienced operating engineer thereafter could bring the cage or skip within an inch of the requested level, even though it might be twenty-five hundred feet or more below his station.

For some unfathomable psychological reason hoist operators very seldom let their equipment go out of control at depth. On the other hand, many accidents were caused by the operator failing to stop his hoist when the bucket or cage came to surface. The Cripple Creek sabotage was of this nature and of course was not the operator's fault. When inadvertently done, such a failure meant that the cage was slammed up hard against its guide stops or even against the sheave itself. This invariably broke the hoist rope (usually where it was connected to the cage, a weak spot in the rope because of vibration, whose energy was absorbed at that point, causing fatigue of the strands), and the cage would then fall the length of the shaft unless it had been provided with some sort of safety brake.[24]

Now and then, however, equipment might go out of control at depth for one reason or another. A Colorado School of Mines student enjoyed such an event at the hands of a friend, who related:

One summer . . . we were earning a few bob surveying a small underground mine in our beloved Colorado. The mine had three levels served by a winze. The bottom level was about 20 ft. above the bottom of the winze. Naturally, the operators—being prudent men—kept the water pumped down to about 3 ft. below the third level, allowing the bottom 17 ft. of the winze to fill with water.

The winze was equipped with an antiquated lump of iron hysterically called a friction hoist. This hoist was made to operate with steam but had been converted to compressed air with the standard conversion kit which consists of two pipe wrenches.

For reasons which we never fully understood, this hoist only ran one way, which was up. It would not run down, and the only way to descend was by crafty manipulation of the clutch and brake, which had only two combinations, these being on and off. In between, gravity took over. . . . Naturally, it took some bit of skill to land a free-swinging bucket somewhere near the level where one wanted to dismount.

In a matter of a day or so, the first level was mapped, and it was time to carry our azimuth to the levels below. Our partner, whose name was Gordon, was prepared to shoulder the gun and descend the winze by ladder. We immediately pointed out that the ladders were old and rotten and as the leader we simply could not risk injuring an associate by the use of dangerous ladders. We assured Gordon that in our experience, the proper manner in which the hard rock man would descend was via the stirrups on the bucket. [that is, standing with one foot on each end of the bail, where it made a right-angle turn to engage the loops on the side of and quite near the top rim of the bucket.] For a while we received a little static about our experience in operating hoists and such, but we assured Gordon that all that was required was for him to hold the bell rope—and as all hard rock men know, one bell [while the hoist is in motion] means stop. Reluctantly, Gordon got on the bucket, riding the stirrups, transit in one hand and holding to the rusty cable [hoist rope] with the other.

We estimate that the first increment of free fall was about 130 ft., somewhere between the second and third level before Gordon found the bell rope. Peering into the blackness we could note the gentle sway of the suspended bucket by the movement of Gordon's miner's lamp. Gordon was shouting something about our ancestry but nonetheless gave us the bell to proceed slowly. The second increment was terminated by a sudden slack in the cable, followed by a loud splash.

In the years that have followed, we often see Gordie at conventions and talk about the old days, but we never stand very close to him if there is a pitcher of ice water within easy reach.[25]

Useful as the bucket was—it was equally good for hoisting rock and bailing water or for lowering men and materials—it had disadvantages which were uncomfortably exaggerated as shaft depth and hoist speed increased. For one thing, buckets were prone to rotate dizzingly under the influence of the twist of the rope. For another, they were inclined to oscillate considerably, banging from one side to another of the shaft unless constantly and laboriously fended off. Additionally, a bucket was difficult to dump when full of rock or water, although exceptionally large bailing buckets were usually provided with bottom valves to facilitate both filling and emptying. Unless quite large they had a distressing

A jib-back surface incline hoist, Sheridan Mine, Colorado.

tendency to be dangerously top-heavy when carrying passengers or materials. Counterbalanced buckets often got their ropes tangled inextricably. From first to last in the old West, buckets remained indispensable in shaft sinking, but the miners' preference for other forms of personal conveyance is a good indication of their status as a necessary evil.

The development of the incline skip, probably in Cornwall or the Welsh collieries, came possibly as a logical hybridization of hoisting and tramming. It would require no great stretch of the imagination to conceive of pulling railed cars to surface up an inclined shaft by means of hoisting gear. Inclines are not easy to sink truly, but with some care in surveying it can be managed. An ore car, held to its rails by flanged wheels and its own weight, would be considerably easier in all respects to raise to surface than was a bucket. In addition, the incline itself would greatly reduce strain on the haulage rope and hoist. In a sufficiently wide incline two such cars could be employed to good advantage in counterbalance. Of course, inclines would be very easy to install in the open air when open-cast mining or moderately steep grades between the mine portal and the mill made conventional ore wagon haulage difficult. In this last connection, the visitation engineer, James V. Thompson, tells a story of great significance:

. . . We would like to record for history the deeds of one of the most brilliant master mechanics the mining industry has ever known. We used to have to visit a miserable open pit mine on top of a mountain. The ore was lowered down the side of the mountain with an ingenious device known as a jib-back hoist [two incline cars in counterbalance]. This machine was not quite perpetual motion, but it was almost something for nothing. The hoist had two drums keyed to a common 8-in. shaft. The ropes were overwinding on one drum and underwinding on the other, and, as a result, the two skips moved in balance in opposite directions when the hoist was in motion. The loaded skip, upon descending the incline, would pull the empty skip back [up] to the ore loadout bin. This hoist was strictly an economy model. There were only two bearings with none in the middle between the drums.

Careful analysis of the stress patterns of this hoist would lead one to believe that it would be subject to frequent failure from shaft breakage, and we can testify that in this case theory and practice were in close accord. The shaft broke about once a month, providing considerable excitement as the empty skip raced back over the mill bins, through the wall of the building and out into the river in a graceful parabolic arc.

Once or twice during a period of 30 years the master mechanic suggested that it would be a good idea to replace the hoist with a better one, but the management gave him the violin music about cash flow, proper use of capital and the tax picture. This always made him feel that he was one of the management team, so he would go to the local machine shop every month and buy a new 8-in. shaft 54 in. long with milled keyway, costing about two and one-half times the book value of the hoist.

Replacing the broken shaft after each thrilling adventure resulted in much down time, and the master mechanic decided that if he could not get a new hoist, he would just have to find a way to reduce down time. His solution to

103

the problem was sheer genius. He reasoned that one shaft 162 in. long would cost less than three shafts 54 in. long and that removing one-half of a shaft took less time than removing both pieces.

On our next visit to the hoist station, 54 in. of shaft was in place between the two bearings, and 108 in. was hanging out in cantilever fashion. Our master mechanic proudly explained that as the shaft broke, he removed the short broken piece and fed in more shaft. Genius like this should not be wasted.[26]

Meanwhile, back at the drawing board similar geniuses were musing that if a skip could be made go up an incline it could, with some slight modification, equally well go up a somewhat steeper shaft or even a shaft with varying degrees of inclination. Therefore they redesigned the skips until they came to resemble elongated boxes (open at the top) in which the wheels had little more duty than merely to guide the skip on its rails. A judicious combination of such wheels on both sides of the skip engaged appropriate guide rails set in whatever portion of the incline was temporarily the "down" side. With rollers between the rails to protect the haulage rope from frictional drag, such a skip would readily follow radically different angles of ascent on its way to grass.

At first, emptying the skip was a problem solved by equipping it with a hinged side or bottom secured by a pin. When the pin was knocked out, the side would open, allowing the ore to fall out. A better arrangement was the self-dumping skip in which the haulage rope was attached near the bottom. When it reached surface, the upper guide wheels followed the rails to outward-curving scrolls securely attached to the headframe. These scrolls guided the mouth of the skip outward over the ore bin and then blocked its forward movement. Continued hoisting raised the bottom of the skip above the level of its mouth, spilling the contents into the ore chute or bin by the power of the hoist instead of by human labor. Possibly the only ineradicable trouble here was safety: there was hardly any way of checking a skip whose rope had broken. In consequence some very nasty accidents took place until it was agreed that men should not ride skips unless there was absolutely no alternative method available to them.[27]

As the ore car evolved into the skip, so the bucket was transformed into the cage, or mine elevator, first made feasible by the use of wire rope and after 1860 customarily installed in shafts after their completion. The first cages were simple platforms suspended from an inverted U hanger to the top center of which the hoist rope was attached. The cages were held in alignment in the hoist compartment of the shaft by rollers or U-shaped shoes sliding along well-greased wooden guides. Wood was and is superior to metal for this purpose for various reasons. One is that, despite all care in installation, the guides cannot be set in exact alignment. The shoe, however, will soon wear away the wood in the areas of friction, eventually providing smooth travel. Such guides are made of clear, straight-grained four-by-six-inch timbers in twenty-foot lengths that can easily be replaced when needed. They have never been cheap, but they pay their way. Should the ground begin to move anywhere in the vicinity of the hoist shaft, the first sign will probably be the cage sticking on its guides, which have been imperceptibly thrown out of alignment; indeed, the first duty of a hoister is to put an empty cage up and down the shaft one trip before the men load up. Some experiment will outline the level and direction of ground movement, whereupon appropriate countermeasures can be taken.

Cages had hardly been introduced when there arose the question of hoist-rope failure and the inevitable consequences. Many patents for cage emergency brakes were taken out in the period from 1862 to 1870, the earliest being based on the principle of a center-hinged, spring-loaded cage suspender, to the hinge of which the hoist rope was shackled. The outward-pointing ends of the suspender arms were equipped with toothed points (or, in later days, linked to deeply serrated cams called "dogs") opposing, but just out of contact with, the sides of the timber guides. The weight of the empty cage was just sufficient to draw the arms down against the tension of the spring, retracting the points or rotating the cams clear of the guides; any additional weight in the cage retracted or rotated them still farther. Should the rope break, the release of tension at the suspender hinge permitted the spring to reverse the cams or force the points outward. Either was so inclined that the weight of the cage forced the points, or serrations, into the wood at a forty-five degree downward angle—the greater the weight of the load, the deeper the "bite" into the wood. Hence the preference for wooden guides, since steel guide rails could not lend themselves to such treatment. Bold inventors were fond of loading their patent cages with the maximum tonnage, seating themselves on the load, and

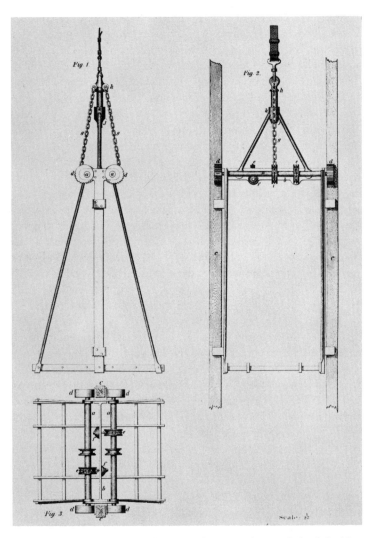

An unenclosed mine cage. The cage shown is almost skeletal, lacking sides or overhead protection. Serrated cams serve as safety-brake dogs. The clevis of the hanger is attached to the end of a braided-wire hoist belt. *C*, the timber guides. From Clarence W. King, *United States Geological Exploration of the Fortieth Parallel.*

bidding their assistants cut or unshackle the rope. To the relief of the assembled spectators the cage would drop only an inch or so before it halted—in most cases.

Since the cage was suspended from its highest point, no danger of overturning existed as with a bucket,[28] but miners still did not care for its total want of overhead or side protection.[29] It is not economical to lag deep shafts all the way to the sump, but in the absence of such sheathing, blocks of rock tended to detach themselves from the shaft sides as a result of vibration, ground movement, or air slaking. Those persons on whose heads these fall are not amused. Presently a sort of gabeled iron roof, the "hood," was installed beneath the shackle and brake to afford a measure of protection against this unwelcome bombardment, but the sides of the cage remained unenclosed for a long time, and when it was in motion, the miners had no recourse but to grip its slender iron stanchions and hang on for dear life. They soon became accustomed to this, but visitors found it difficult to affect unconcern at what was nearly free-fall descent, plus the spectacle of rough rock walls flashing past six inches from their unprotected noses. Complicating their sensations was a more or less nauseating vertical oscillation below one thousand feet, produced by the stretching and contracting of the elastic steel rope as the hoist engineer far above eased on his brake.

This habit of hoist ropes stretching at depth proved occasionally embarrassing to cagers, who found that a loaded ore car's weight depressed the cage from six inches to one foot as the front wheels rolled onto the cage's stub tracks from the level turnsheet. The sudden stretch of the rope would tilt the car alarmingly, spill its contents, or even overturn it. To correct this tendency, "chairs" were devised, these being a variety of devices to lock the cage temporarily in place while loading. They ranged from lever-actuated sliding bars like door mortise locks to short lengths of chain slipped under hooks mounted on the cage. All worked well—too well, in fact. Cagers never forgot to engage the chairs but very frequently neglected to disengage them before ringing one bell. Then the hoist shaft would be so filled with the profanity of God-fearing workingmen that, it is said, the language would have to be shot out and barred down at considerable expense to the management. At length an automatic chair was invented, bringing peace and satisfaction to all parties but the devil.

Once high-speed hoisting from great depth was made safe and economical, American mining in the 1870's assumed a pattern which would be retained in typical developments almost to the present day. The ore body was first developed by sinking the hoist shaft on whichever side of the contact between lode and country rock seemed to give the greatest promise of good ground. From this shaft levels would be driven laterally into the ore body at (approximately) one-hundred-foot vertical intervals. Each level would be gradually expanded by overhand stoping, with the ore shot down being mucked into ore cars trammed by men or mules to the level station. The cage was then summoned, and its attendant "cager" would push the cars onto stub tracks secured to its floor for hoisting to grass. Under other conditions the ore would be mucked into wheelbarrows and spilled down transfer raises into chutes which terminated in a "pocket" at the skip incline. Here the skip tender would pull open the gate to fill the skip on arrival.

A prophetic improvement on this system was employed on the Comstock Lode about 1875. The heart of the arrangement was a very deep tunnel running beneath the ore body and inclined to match the dip of the footwall in which it was constructed and which afforded good ground. Tunnel and footwall were about fifteen hundred feet below surface on the west (where the "third-line" hoist shafts intersected them) and ran east at about a thirty-degree inclination below the horizontal for some hundreds of feet. When this inclined tunnel was completed, ore from the stopes above was to be dumped down chutes which had mouths terminating therein. An extra-large five-ton capacity skip or tramcar was installed in the tunnel. This car was nicknamed the "giraffe," since its front (western) wheels were considerably smaller than its rear wheels, enabling the box to sit approximately level despite the inclination of the haulage way. The ore was run into it from the chutes, and the giraffe was then pulled to the hoist shaft with a rope wound by a steam hoist at surface. Observers were careful to note that the rope ran *under* the sheave set in the bottom of the shaft, as of course it had to do. At the foot of the shaft the giraffe unloaded into two-ton Comstock-pattern ore cars, which were then conventionally raised to surface in a three-decked cage.[30]

In some respects this arrangement crudely anticipated modern drift and stope slushing. Slushing employs an electric or air-driven winch, some sheaves and rope, and a dragline scoop which can be pulled back

Cagers loading cars into the cage. Note that the level station is floored with sheet iron (the turnsheet) to allow maneuvering of the cars. The bell rope is at the right of the cage. Reproduced by permission of Buck O'Donnell and Shaft and Development Machines, Inc.

and forth along the floor of the haulageway. When a sufficient amount of ore has been shot down onto the floor, the empty scoop rattles and bangs over the muck pile, halts at the operator's pleasure, seizes a large, untidy mouthful, and drags it in the reverse direction to the mouth of a transfer raise, or chute. The ore vanishes down this ominous maw, the winch is reversed, the empty rakelike bucket returns to the far end of the line of muck, and a new cycle begins. Like the Comstock giraffe, this provides for swift horizontal transportation of ore by mechanical power, it is actuated by a rope, and it completely eliminates expensive hand mucking.

The discovery of the great telluric gold lodes of the Sawatch Rockies in the San Juan District of southwestern Colorado in the late 1880's demanded yet another system of ore transportation. For geological reasons satisfactory to themselves, these lodes tend to crop out halfway or more up the slopes of very steep, high mountains. Once the prospectors caught on to this peculiarity, they located them by panning up the valley stream beds to the point where the gold colors abruptly faded or quit. Backing down a bit, the prospectors then turned themselves into Alpinists, clambering and scrambling upward until they found the outcrop of origin, which might be a thousand or more feet above the valley floor. The concentrators had to be built beside the streams in order to get the water they required, but the portal of the mine was like the moon—in plain sight above but difficult to get to. Ore-wagon transportation was totally out of the question; however difficult it may be to get a loaded wagon up such a slope, it is a hundred times more troublesome to get it and its load safely *down*. On the other hand, the slopes were not quite steep enough to use chutes with any prospect of success.

These famous (and today reviving) mines—the Smuggler Union, Camp Bird, Bella Union, Good Hope, and so on—therefore installed aerial tramways, very like modern ski lifts in conception save that the bulk of the payload was downward bound. Substantial wooden pylons were erected on the slope, equipped with patent sheaves, and rigged with an endless wire rope whose upper bight was run several times around a steam hoist drum and tension-adjustment machinery at the portal. Suspended buckets, the largest capable of holding a quarter ton, were spaced on the rope as indicated, to be loaded at the portal ore bin and dumped at the mill. Since by far the greatest part of the tonnage was assisted along by gravity, optimists felt that the tramways generated so much power that crushers might with profit be linked into the moving line to take advantage of this fact. Some experiment showed that their hope was unfounded, since the surplus energy was so erratic and undependable as to be useless. Indeed, the portal hoist engine was essential both as a brake and to keep the line moving at a uniform rate of speed—the erratic movement of an uncontrolled line set the buckets swinging and spilling their loads, racked the pylons, and set up destructive vibrations in the rope itself.

Those travelers who today go from Cortez to Telluride or from

Durango to Ouray may observe the remains of such tramways scattered through the mountains. In the days of their flowering they were imposing spectacles. The North Star Mine tramway, for example, had an over-all length of two miles, in the course of which it descended thirty-two hundred feet to the mill on the Animas River. The line was supported by fifty pylons and was equipped with forty buckets, each capable of holding six hundred pounds. Men were strictly forbidden to ride these buckets, but when deep snow isolated the mine for weeks on end and thirst and other appetites became overpowering, the buckets, moving in stately procession, afforded the only means of obtaining whisky and the domestic consolations. Desperate situations imply desperate remedies, and even professional engineers were known to flout their own rules in this respect when pressed.[31]

In the western mines horizontal tramming was done with one-ton iron ore cars, sixteen of which constituted a mucker's quota during his shift. The cars were equipped with four closely set flanged wheels which ran on eighteen-inch-gauge track. Their boxes were so suspended that on tripping a latch lever they dumped sideways, but many had a center pintle so that if necessary the box could be swung ninety degrees and dumped end-o. One loaded car was about all that a man could push, but if mule haulage was used in the mine, trains of up to ten cars could be linked together with short chains with hooks dropped into corner eyes. The mule's singletree was connected to the leading car by two short trace chains, which joined in the center of the leading car to form a V on which the driver could seat himself as though in an exiguous hammock. Although the seat of his trousers nearly scraped the floor of the drift, his head would clear by a wide margin any overhead obstruction.

The tracks were composed of lengths of light iron rails secured with miniature track spikes to the small crossties. In large, busy mines the tramming levels were often double-tracked and had turnouts and junctions equipped with hand-operated track switches. It has been said (for what it is worth) that a veteran mule coming up on a track switch set the wrong way would halt long enough to bat it into the correct position with a tap of his hoof. This legend leads to the observation that, in modern drift mines where "electric mule" tramming is still used, red and green block signals are installed along the track, just as on a surface railroad. The signals are square lanterns in which panes of colored glass appear to be irresistible targets of opportunity to those miners of artistic bent who desire a medium for graphic self-expression. After one such tour the author concluded that the prejudice against admitting females to drift mines may be in some part based on the disinclination of the management either to have the ladies' sensibilities affronted or, alternatively, to spending really considerable amounts of money effacing the products of years of devotion to the muses of classic art and literature.

Once they had been hoisted to grass, ore and waste were moved about the mine yard in the same cars which had been loaded below. Waste was spilled over the edge of the mine dump, and ore was taken to the shack, in which it was sorted and upgraded before going to the ore bins. From these bins, set over the haulage road, it was tapped down into the narrow, high-sided ore wagons (most familiar to the public in connection with Death Valley borax transportation), which carried it to the mill. Once in while, if water was in plentiful supply and gradients favorable, it was economic to crush and stamp the ore at the mine head and then pipe it as a thin slurry down to the mill, but this practice was comparatively rare.

In the Comstock Lode, sometimes called "America's first school of practical mining," circumstances virtually forced on the engineers the conclusion that large tonnages could be moved most economically by force of gravity. From the lowest point of the mine the ore should then be hoisted by machinery. One Irish mucker received a seemingly paltry daily wage, but on examination it was shown to amount to $0.225 a ton for the relatively simple labor of lifting ore four feet during the initial step of the process. This was shockingly expensive when contrasted with the costs of a chute or even a large steam hoist. The growing tendencies of the miners to unionization provided another pressing motive to economize on manpower. High costs might be acceptable in bonanza ore, but bonanzas were growing few and far between. The day was coming increasingly near when out of sheer necessity the ore would have to be moved its entire course from stope or bench to smelter mould totally untouched by human hands.

CHAPTER VI

Hidden Gold

IN theory gold should be one of the easiest metallic minerals to prospect. Gold-quartz lodes are often marked by iron and manganese whose weathered stains make the country appear "burned" or black and the lode outcrop rusty red. Prospect pits sunk to bedrock down the lines of wash from such formations will reveal granules of gold colors which are easily recovered by panning or even air winnowing. Quartz is there or thereabouts, appearing as intrusive white threadlets, snow-white outcrops, white stream pebbles, or the cindery "gossan," which is spongy and mineral-stained. Often enough, gold-bearing intrusives may assume gross features which under favorable conditions may be observed for miles across open country. Where old, black shales and schists suddenly give way to recent granites, the abrupt change hints at major earth faults favorable to the deposit of gold ore at or adjoining this discontinuity. And in historical fact gold is usually the first mineral to be found in favorable mining country, even if it should later prove that much larger deposits of different metallic ores will become the economic mainstay of the district.

Now and again, however, expectations fail to match up to the facts. The prospector's signposts may be absent in whole or in part or appear in a manner not at all what tradition assumes. In these cases the truly gifted prospector comes into his kingdom. One long look through his field glasses was enough to start Ed Schieffelin on his successful search for bonanza silver in a region which had been thoroughly raked over—and by some pretty good men at that—for three centuries. Like Schieffelin, a few intelligent and persevering desert rats were able to fasten on one indication of mineralization and track it relentlessly through a morass of deceptiveness until they hit pay ore. Some became wealthy, and some died in abject poverty—such is luck—but their achievements deserve to be recalled as shining examples of what a really good man can do when he sets his mind to the task at hand.

Before 1862 no gold worth going after was thought to exist in Arizona Territory. The Spanish, Mexicans, and Americans in succession found it best to stay south of the Gila River, working the silver deposits, most of which were around the Santa Rita Mountains, no great distance from Tucson. North of the Gila the land was regarded as Cain's country, uninviting in itself and occupied by Apache Indians, whose enterprising ways with intruders were proverbial. In 1858, however, the unsavory "Colonel" Jacob Snively, who had made even tolerant Texas too hot to hold him, led in a party that found placer gold in the mouths of the big dry washes falling into the Colorado River from the east, twenty miles east of Yuma Crossing. This strike produced the Colorado River excitement which gave rise in 1862 to the camp La Paz. One year later the scout Joseph Reddeford Walker led a party up the Hassayampa River to the Bradshaw Mountains, more or less central to the territory. In the pediment watercourses and gulches more color was found, inciting a rush to Poland Camp and Lynx Creek.[1] In neither case were the prospectors able to locate immediately the lodes from which the colors had originated, since, as it turned out, they were small, scattered, and fairly well concealed.

Some time in 1863 the Austrian-born prospector Henry Wickenburg entered the territory with the Pauline Weaver party. Born in 1820, he had emigrated to the United States in 1847 and had gone to California in 1853. There he learned to prospect and had been attracted to Arizona by the La Paz and Bradshaw excitements. He accumulated a stake by working for a time as a government teamster near Tucson. In late spring or early summer of 1863 he joined a party led by Erving A. Van Bibber heading for the Bradshaws and Harqua Hala mountains by way of the Hassayampa River trail.[2] At the Bradshaw camps he found nothing of interest, and in October[3] he returned south by the same trail, riding for mutual protection with two companions, one named Bloomfield and another whose name has been lost.[4] They descended the abrupt southwestern escarpment of the Bradshaws and struck off south across the flat at its feet. The party was very short of water, and their first objective was the Hassayampa, thirty or forty miles distant at the foot of the gently rolling but completely waterless plain.[5]

It is possible that gold seekers by the score had passed that way before Wickenburg came back, and it is certain that all of them had ignored the first and most elementary rule of their art, which is to scan the country and the skyline continually as they move, looking always for anything out of the ordinary. Not Wickenburg, who evidently began digesting something which he could hardly avoid seeing. The plain he was traversing is broken on the west by low hogback ridges and basaltic flows, except for one unique formation behind them and somewhat on the right (southwest) of the party's line of advance. The formation resembled a group of stubby columns closely spaced; with the exercise of some imagination they looked like the fingertips of a gargantuan hand gesturing toward the sky. The rock of this formation was light red, differing markedly from the duns and cindery blacks of the other and closer elevations. It was impossible to make out more than that from a distance, but the duty of the prospector is to investigate anything which differs significantly from the country in color, contour, or vegetative cover. It was later to be suggested that Wickenburg dowsed with a burro,[6] but that is out of the question—the fingerlike prominences, subsequently named the Vulture Mountains, virtually shouted an invitation to be looked over.[7]

After refreshing themselves at the Hassayampa, the party rode west by north to the formation, a distance of twelve to fourteen miles. Bloomfield and the third man may have been downcast at their first examination of the red rock, for it was an andesitic and rhyolitic lava, the columns being obviously the eroded remains of several eruptive stocks and their associated winglike dikes originating from a common vent. It held no values. Values are very seldom found directly within such formations but are posteruptive features most often found some distance away at the contact between the igneous intrusion and the original country rock. Not that Wickenburg necessarily knew this, although he sensed that something was in the wind. Since the terrain there is cut up by washes and swells, Wickenburg then probably began to cut a wide circle around the necks. Northeast of them he presently found what he was looking for: an outcrop of slate and quartz large enough to be called a butte. It was 100 feet high, 500 feet long, and 250 feet wide.[8] Its black color and its complexity of structure were promising of mineralization.

Since Wickenburg was equipped for placering, it is likely that he merely broke off some surficial rusty quartz, pounded it to powder between two rocks, and then panned it out with water from his canteen. When he had finished, there was a small tail of gold colors in the drag of his pan. The sparse color was disheartening, for the greatest proportional values of gold in lodes are usually at grass roots, where weathering has concentrated them. On the other hand, the very size of the butte was astonishing, and Wickenburg began to get the idea that the gold persisted through all of it. To offset this, even though there might be a great deal of gold in the aggregate in sight, it could not be won without water, and the closest water of any sort was back at the Hassayampa. After more examination and discussion, Bloomfield and the third man remained unimpressed. They rode away, leaving Wickenburg in full enjoyment of his find, which he located as the Vulture. If one discounts unrecorded Spanish or Mexican strikes, it was the first hard-rock gold strike to be made in the desert Southwest and physically perhaps the largest ever.

In the months that followed, Wickenburg became better acquainted with his prospect. The hanging wall was the same red porphyritic lava that composed the Vulture Mountains. The footwall and country rock, as well as the gangue of the lode itself, was a talcose slate dipping at a forty-five-degree angle and striking roughly northwest. After clearing off the scree, he found four quartz leads within the butte, paralleling its dip and strike; evidently the mineralizing solutions had risen between the natural joints of the slate strata. The four leads had a total width of 40 feet and in the best spots contained about 4.5 ounces of gold to the ton.[9] In a later period the comparatively low tenor of the ore would be considered less significant than the very substantial tonnage in sight (about 656,000 cubic yards of pay quartz, based on persistence of the leads to a depth of 245 feet), but at times Wickenburg may have considered renaming his location the White Elephant. Only an ample supply of water and a large, efficient concentrator could make the Vulture pay, and obtaining either was clearly beyond his means.

116

Promising country: hematite and limonite-stained peaks of the San Juan Rockies in the country around Telluride and Silverton, Colorado.

The headframe of the Florence Mine. Goldfield, Nevada.

The adobe cabin of the size and, by repute, on or near the site of that occupied by the Schieffelins and Richard Gird, hard by the Brunckow Mine.

He stayed with it, however, for a number of reasons, primarily because prospectors of the period nursed the conviction (usually incorrect) that ore tenor sometimes dramatically improved with increasing depth. Hence the poor surface showing might not be as bad as it seemed. With care and a great deal of hard work Wickenburg could eke out a living on the Vulture while developing it enough to attract a purchaser who had the capital to exploit it properly. Certainly the size of the outcrop was its own best advertisement. It was far larger than anything then known in California, while even the footwall and the slate strata between the quartz leads contained values of gold carried by pyrites. Finally, the element of Teutonic stubbornness should not be discounted, nor the belief that a low-grade lode in the hand was worth any number of bonanzas hidden behind the hills. At any rate, Wickenburg would dwell ever after within sight of the red skyward-pointing fingers of the Vulture Mountains until his death by his own hand in 1905.

When confronted with such a situation, the western prospector of the old school fell back on standard operating procedure. He hunted for the best-looking spots of ore and broke them out with a quartz pick or miner's bar. The ore so obtained would be sledged down to fist size and carefully sorted over for pieces of the highest grade. Two hundred pounds of this selected and upgraded ore could be carried by a pack animal the twelve miles to the Hassayampa River. In this hot and onerous work Wickenburg was assisted by James A. Moore, a drifter whom he persuaded to go in with him on a promise of equal shares in the eventual cleanup.[10]

In the spring of 1864, selecting a suitable site near the bank of the river, Wickenburg collected boulders and began to build an arrastra, a primitive ore mill whose design had been brought to the Americas by the Spaniards. It consisted of a circular floor of flat-surfaced boulders set in well-puddled clay and secured by wooden wedges and more clay. A low stone coping ran about the periphery of the floor. Pivoted stoutly over the center was a wooden sweep from which two drag stones were hung so as to rub heavily over the floor. One end of the sweep was prolonged outside the coping sufficiently to permit harnessing a blindfolded horse or mule thereto. The floor was covered with an inch or two of well-broken ore, and the animal was urged to

A Mexican mill, or arrastra. The mill is very primitive, having only one drag stone. The animal should have been blindfolded. Henry Wickenburg's arrastra was probably no better than this. From Louis Simonin, *La vie souterrane*.

plod around the arrastra, the sweep pulling the drag stones over the ore to grind it. From time to time the operator poured in some water to assist the process. After a day or two the ore was reduced to sand or flour, freeing the gold particles from the gangue. Then the floor was carefully flushed with water to wash most of the floured tails out a scupper. With a small horn spoon the operator now scraped up the mixture of gold particles, black sand, and silt from between the floor stones. He panned it out at the water's edge, dried his cleanup, puffed it free of silt, and added it to his tin cup or other container.

On July 4, 1864, as Wickenburg was about to put his mill on line, a Peeples Valley rancher, Charles B. Genung, rode up the trail from the Gila and, seeing Wickenburg's "poor excuse for an arrastra," lent a hand in making it better.[11] Genung hung around during the first run, whose cleanup netted 14.5 ounces of gold. This at first seems improbably high for Vulture ore, but it must be remembered that only selected pieces had been milled thus far. Nevertheless, at the frontier price of $15 an ounce, this amounted to but $110 for each partner, a poor recompense for months of hard work. After viewing his share, James Moore quit.

Wickenburg was unable to work the Vulture alone. He decided to lease it by selling the ore in place at fifteen dollars a ton, leaving it up to the lessee to do his own quarrying, transportation, and milling. Rather surprisingly, he attracted quite a number of takers, few of whom appear on the evidence to have had much experience with the economics of gold mines. They seem to have had the impression that thirty dollars in ore would net fifteen dollars' profit after Wickenburg had been paid. They also fancied that open-cast quarrying in the open air would be easier and cheaper than the drift mining which was then beginning to get under way in the newly discovered Bradshaw quartz lodes. They may also have hoped that they themselves might hit a rich pocket of jewelry ore or a high-tenor "hot lead" in an ore shoot. Consequently, there was a gathering of speculators in late 1864 and early 1865 along the river bank at the camp they would call Wickenburg, including among their number the father of Phoenix, Jack Swilling. It would be no great time until they were educated to the fact that even a gold mine has to show a profit to stay in business.

Within the year forty arrastras were at work on Vulture ore along the Hassayampa, and a few operators were arranging to freight in and erect stamp mills of modest capacity. On the basis of aggregate cleanup, the lease operators appeared to be making a great deal of money, but some analysis of their costs tells a different story even in respect to Wickenburg, who seemed to have nothing to do but sit in the shade and count the $15 he supposedly collected on each ton that left the Vulture. Despite the supposition that gold mining is a sure way of getting something for nothing, ore haulage to the Hassayampa alone was costing the lessees $17 a ton. Mining cost $2.50 and, while their milling costs cannot be estimated, inefficiency in their crude milling was losing them $5 a ton in their tailing. On mine-run $30 ore they were marching swiftly to financial ruin. Wickenburg himself had to fight the usual expensive lawsuit over ownership of the property, and it is more than likely that he was forced to defer royalty payments from his lessees.[12]

After most of the lease operators went under, in 1867 a Philadelphia promoter, Behtchuel Phelps, organized the Vulture Company.[13] It bought four-fifths of the location footage from Wickenburg for $85,000, most of which he agreed to forgo for the moment. Charles B. Genung was put in charge at the mine yard,[14] while a forty-stamp mill was installed at the river under the management of Phelps's brother-in-law, Ben Sexton—the name would prove sadly appropriate, as events would show. For the next six years the Vulture Company flew into the butte with energy, its 102 miners reducing the elevation to a glory hole. The mill sold $1.8 million in gold bullion or pyritic concentrates, while impounding six thousand tons of middling concentrates and twenty thousand tons of tailing for reworking when transportation became cheaper.

Mining men and other grandees inspected the Vulture, whose fame exceeded its values. J. Ross Browne and General James E. Rusling visited the development in 1868, and Rossiter W. Raymond was shown around in 1870. All three were much impressed by the extent of the lode. Browne's and Raymond's reports were virtually identical, both men probably being given the same tour and the same statistics by Genung. Their host was candid about the ore values, a consistent thirty dollars in the quartz, tonnages broken out and milled, and other matters of moment. Both visitors decided to let the figures speak for themselves—as matters stood the Vulture was no bonanza, but, given railroad transportation, its profits should increase considerably.[15] In passing, Raymond noted that Wickenburg was "living now, if not in needy circumstances, at least provided with less than an average share of worldly goods, near the town of Wickenburg, where he owns a small farm."[16]

If all was not well with Wickenburg, neither were things well for the Vulture Company. Hoping without good reason that the lode would become richer at depth, the company sank two inclined shafts on the ore, hoisting from the 245-foot level with horse whims. It found that the ore horizons developed were no better than at the surface, thus wasting the considerable investment in this work as opposed to the much cheaper open-cast mining. In fact, the deep ore carried appreciable percentages of copper, galena, and zinc, weathered out of the oxidized zone above. These minerals made concentration more difficult and were not so rich as to have any particular value. Even water (normally regarded by miners as a curse but in this land as a blessing) had not been struck. The last blow was the discovery in July, 1872, that the ore at maximum depth was thinning appreciably in value. In view of the over-all cost of $14.93 a ton to hoist and mill the better ore, it was time for a critical review of the entire operation.[17]

A critical review was about the last thing that Sexton, superintendent of the Vulture Company mill, desired. For six personally profitable years he had been stealing mill concentrates and cleanup amalgam, apparently making off with everything but the cast-iron stamp shoes. He had contracted a reported one hundred thousand dollars in company debt, probably as a device to conceal temporarily his embezzlements. At the prospect of an audit Sexton departed between days.[18] Finding itself penniless, the Vulture Company went into liquidation. Wickenburg's promised payment was included in the debts written off, leaving him in no better circumstances than he had been at the moment his eyes first rested on the fingertips of the Vulture Mountains. It would seem, on reflection, that a Sexton had buried a Vulture.[19]

Before 1870 the Pikes Peak District of Colorado would have seemed as unpromising of profitable mineralization as any place one could find in the Rocky Mountains. To be sure, in 1859 there had been a big excitement one hundred miles farther north in the Front Range when a wandering Georgian, John H. Gregory, had located a major gold-quartz lode at the future site of Central City. Afterward, however, the Colorado boomers had been drawn to the west and southwest, following the Leadville and San Juan prospects. They explored a bit directly south but found little except some slight traces of color in the streams. This meant little—for that matter there is float gold in some streams of Indiana, of all places, but no one is getting rich on it. The metropolis of the district was Colorado Springs, a resort and spa, while the grassy parks on the pediment of the great peak were taken over by cattlemen for what they could make of it, which was not very much.

In view of the country formations, this seemed a reasonable outcome. The country rock of Pikes Peak and its pediment is a very ancient red granite, formed at considerable depth. It was forced upward during the mountain-building Laramide revolution, about seventy-five million years ago. Pikes Peak granite is normally without intrinsic or intrusive values. At a somewhat later time volcanic action erupted through a pediment plateau southwest a few miles from the summit of the peak itself. Nine closely spaced vents blasted out a conical bowl two miles wide, discharging a variety of feldspar lavas (latite, phonolite, and trachyte) whose water and gas contents foamed off rapidly, producing blocks of material physically (but not chemically) similar to pumice. These blocks and the particulate "ash" built a wide, low volcanic cone. Since the fluffy materials were poorly bonded, most of the cone was swiftly eroded and removed, although the loose material over the central bowl settled downward to pack it with a plug of moderately compacted feldspathic breccia. This latite-phonolite breccia mass, surrounding islands of country granite or one later, also brecciated, neck of basalt, carried no more included values than did the Pikes Peak granite. No western prospector would ordinarily have wasted a second glance at any of them.

However, much hydrothermal activity followed the eruption. Since the tight country rock was an impermeable seal, everything had to come up through the breccia plug, whose surface probably vented steam and mineral-spring water for centuries. This water had originated at great depth and had infiltrated fissured and mineralized rock, dissolving from it and carrying up some unusual minerals. Among these were fluorite, pyrite, and vanadium mica (roscoelite), together with high relative percentages of a remarkable mineral, calaverite. As heat and pressure dropped off sharply within the upper fissures of the breccia plug, these values were precipitated and deposited in high concentration. Conforming to the fissures, they formed a series of radially distributed thin, almost vertical silicified lodes within the plug. In a few places on the southeast the mineralized waters followed the contact between plug and granite footwall, shallowly penetrating the granite and replacing its red feldspar with quartz, pyrite, and calaverite. Although the spongy breccia eroded swiftly, the bladelike intrusions were not much more resistant, and, where they protruded slightly from the mean surface, were hidden by outwash scree.[20] Iron stain, a normal signpost caused by weathering down of the pyrites, was camouflaged by the brown color of the phonolatite. The fluffy breccia was so featherlike as to mask equally well the mass of any heavy metal values finely dispersed within it. About the only sign of unusual activity was the purple of fluorite, which at that time was not highly regarded as a prospector's sign.

Calaverite, first identified in and named after Calaveras ("Jumping Frog") County, California, is one of the telluric salts of gold. Tellurium is a semimetal of the oxygen-sulphur-selenium group of elements, with one or more of which most metals will combine. Metallic oxides and sulphides are the common stock-in-trade of the miner, and selenides might be so as well but for selenium's rarity. However, none of this group but tellurium combines chemically with gold in nature; why this is so is something of a mystery. Yet the fact remains that if gold is not found as a telluride it will otherwise be deposited in native particles.[21] The rule-of-thumb prospector was not often familiar with calaverite or its gold-silver equivalent, sylvanite, and usually mistook its shining silvery glint for silver-lead ore. In placers, weathered grains of calaverite darken to "black gold" which will not amalgamate and, when trapped by their great specific gravity in gold-saving equipment, might be readily mistaken for black-iron particles and so be tossed away during the cleanup. Unless a prospector had reason to suppose that calaverite was present, he would neglect to make the simple field test for it and, accordingly, falsely conclude that the few grains of native gold dust (reduced from the telluride by extended weathering) was color so sparse as not to be worth further investigation.

The plateau, whose southwestern quadrant was traversed by a deeply cut wash named Cripple Creek (from its way with the legs of range cattle attempting to cross it) therefore constituted a geological puzzle of labyrinthine intricacy, even to a seasoned prospector, which Robert "Crazy Bob" Womack definitely was not. He had moved into the Cripple Creek basin about 1873, intending to ranch there in a small way, but very soon after became obsessed that gold was there or there-abouts.[22] By one account he had heard that a party of United States Geological Survey cartographers, mapping the region, had come across favorable indications in the gray trachyte.[23] Another version has it that his cousin, Theodore H. Lowe, had found gold float while homesteading the basin range.[24] A small syndicate headed by H. T. Wood (supposedly of the USGS party) and Benjamin Requa had driven a one-hundred-foot adit into the basalt breccia neck near the center of the district but had struck only a mineral which they deemed "white iron" and therefore called it deep enough.[25] Panning by Womack revealed gold colors in Cripple Creek. His idée fixe grew with the passing of time and the discovery in May, 1878, of a piece of trachyte which assayed two hundred dollars a ton, but Womack could find neither its point of origin nor a financial backer. Assayers snorted at the idea that feldspathic lava could carry values, while potential grubstakers asked for better evidence than Womack could produce.

All seemed lost when in 1884 the infamous Chicken Bill Lovell, using the money he had cozened from H. A. W. Tabor by the Chrysolite salting caper, moved into the Pikes Peak region to repeat his fraud. Selecting a phonolite peak, Mount McIntyre (mistakenly confused with Mount Pisgah), somewhat west of the Cripple Creek District, he and some cronies dug prospect pits to bedrock, salted them with gold dust, and then talked up a strike in the saloons of Leadville. An excitement followed, lasting only the four days necessary for the boomers to discover the plain and rather silly truth. Lovell's profits are not known, but the boomers went away convinced that phonolite was worthless as a carrier of gold—and so it is, excepting subsequent mineralization of an entirely independent nature—and that the entire region was *borrasca* (played out) ground.[26]

Unaffected by this event, Womack at last obtained a small grubstake and on October 20, 1890, sank a prospect pit hard by a phonolite outcrop which was outside the main breccia plug. He found samples which assayed up to $250, located his prospect as the El Paso, and optioned it to the speculator F. F. Frisbee for $5,000.[27] It would seem that Womack drew no specific conclusions from his strike but merely continued his rather aimless poking about. The very nature of the El Paso gold values was unknown until, it is said, a newly certified assayer, Ed de la Vergne, saw some of Womack's ore samples on display in Colorado Springs and was struck by their resemblance to sylvanite, with which he was acquainted. He reputedly made the rough field test for a telluride mineral by placing Cripple Creek float of his own finding atop a hot kitchen range. When heated sufficiently, the tellurium combines with atmospheric oxygen to fume off as a white vapor. At the same time the reduced gold emerges from the surface of the sample as a yellow blister. De la Vergne obtained a positive reaction. Now that the nature of the Cripple Creek values had been established, all that any prospector had to do was place a promising sample in the coals of his campfire. If he got his white puff and his metallic yellow blister, he had found calaverite.[28]

Since a concurrent rush to Creede, Colorado, was drawing away boomers who might otherwise have been attracted to Cripple Creek by these discoveries, the district did not really get under way until the arrival in 1891 of the carpenter-prospector Winfield Scott Stratton. Stratton, in contrast to Womack, not only was an experienced prospector but had continued his education by reading geology and learning the elements of blowpipe field assaying. Accompanied by an associate, Dick Houghton, he entered the region early in the year, intending to prospect it for cryolite, a mineral then much in demand as essential to the new Hall process of electrolytic reduction of bauxite to metallic aluminum. Getting wind of the local interest in gold, the partners unpacked their pans and began to search for color somewhat east of the El Paso location. A rancher, Billy Fernay, observed them at work and helpfully guided them to a slope of ground on the lower reaches of which he had seen float—none realized it then, but midway up the slope was the country granite rim and contact zone. Finding color at grass roots, Stratton traced it upward to the base of the bold outcropping granite "comb," at whose foot a lode or mineralized contact zone ought by rights to be found. Trenching along this base came to nothing—the float went right to the granite and then stopped. Stratton was inclined to dismiss the outcrop itself despite its suggestively brown stain, for granite was notoriously without values unless infiltrated by the quartz stringers that carried mineralization. Nothing resembling such stringers, and no other favorable sign seemed apparent to him.

While Stratton was musing over the illogicality of the situation on the evening of July 3, 1891, the penny suddenly dropped. The origin of the float color had to be in the granite itself, appearances to the contrary notwithstanding. He rose at dawn, rode to the spot, and pegged the Independence and the Washington locations on the comb.[29] Samples of the granite went nineteen ounces of gold to the ton ($380 at then-current prices). Stratton had struck the most heavily mineralized segment of the entire circular granite footwall which surrounded the volcanic breccia; later exploration indicated that his locations were almost directly above one of the nine vents through which the mineralizing waters had risen. The iron oxides with which the outcrop was stained should have been his clue, but Stratton did not reproach himself on this account. He later optioned the Independence, regretted his decision, and managed to reclaim it. Seven years of development netted him $2 million, after which he sold the location to a British syndicate for $10 million.[30]

With high values and tonnages proved in the footwall, the Cripple Creek excitement of 1892 began. Rather to his own surprise, Stratton (unusually, for a prospector) proved to be an able promoter and consolidator with a flair for overreaching the host of pirates who attempted to seize some portion of his properties. His Portland Gold Mining Company annexed virtually all of the profitable ground along the southeastern footwall and then joined the rush for the radial lodes in the center of the breccia plug. Prospecting for these was conducted in an unusual manner: the scree was simply cleared away by mule-powered road scrapers, and then locations were pegged on any silicified areas marked by the light-purple fluorite that came to be considered the best indication of a profitable lode. He was not immune to mistakes, but his luck invariably saw him through. For example, rather than pay through the nose for custom milling, Stratton established his own mill, which depended on conventional stamping and concentration followed by amalgamation in pans and tables. Since calaverite will not amalgamate, mill recovery was only 18 per cent of assay values. He soon sensibly gave up the project and shipped his concentrates to Pueblo for preliminary roasting in the presence of oxygen to reduce the tellurides, followed by conventional smelting to clean out the iron, quartz, and other undesirables.[31]

Death itself probably saved Stratton from a worse blunder. Toward the end of his career he conceived a theory that the Cripple Creek values had originated from one single vent more or less centered below the breccia plug. There was no harm in this theory, but there was considerable potential for loss in his next assumption that a very deep shaft sunk over that point would at great depth strike a huge deposit of

126

values, of which the radial lodes and footwall mineralization were but minor branchlike extensions. This was a manifestation of the old "the deeper, the richer" theory so long nourished by frontier mining men and so wasteful of money and so unproductive of results. Scientifically trained engineers knew better, but Stratton did not lack determination, and it is likely that he would have thrown away a good deal of his fortune on the project had he been preserved.

The story of Cripple Creek development needs only cursory summarization. There was a great deal of high-grading in this camp and its twin across the surface divide, Victor. Both enjoyed the usual meed of labor troubles which struck western mining around 1900. The lodes were worked in a fairly conventional manner by sinking a hoist shaft and a ventilation shaft through the breccia to an appropriate depth, running levels therefrom to hit the ore bodies, and then stoping overhand to move the ore down chutes to the tramming levels. These stopes were for the most part high, narrow, and somewhat erratic, timbered by horizontal stulls which provided a convenient base for the working platform. Not infrequently two or three mining syndicates would share shafts, so closely were the radial lodes spaced and so comparatively small the tonnage hoisted. The total gross production of the district to 1902, the end of its flush times, is estimated at $125 million; it could have been more and the working life of the district greatly prolonged but for the uneconomical practice of going after the richest ore and ignoring anything less than bonanza—"picking out the eyes of the mine," the Cousins called it.

Water was struck at about the eight-hundred-foot level below the mean surface. Some pumping was attempted, but it soon became apparent that drainage adits driven beneath the entire district would give better service, since their portals would be substantially lower than the plateau on which the mining was conducted. This water was both cold and stagnant, possibly a relict of the hydrothermal period, and once drained would be gone for good. The chief drainage adit was the three-mile Roosevelt Tunnel, completed and successful in 1916,[32] although exploratory workings below still required pumping to elevate their water to its drainway.

Finally, there was the affair of the Cresson vug, perhaps the greatest high-grade gold deposit ever found in any room-sized area outside the vault of a national treasury. The basalt breccia pipe, the site of the original Wood-Requa adit and north a mile from Stratton's footwall discovery, had during the boom been relocated as the Cresson. It was worked none too profitably by several sets of owners, and in 1894 it passed into the possession of two Chicago high rollers, J. R. and Eugene Harbeck. They hired Richard Roelofs as superintendent, making an excellent bargain, for Roelofs was an able engineer with an inventive turn of mind. He stoped fifteen-dollar ore so efficiently that the Cresson paid dividends for the next twenty years, continuing in operation long after most other Cripple Creek mines, haphazardly gutted of their best ore, had shut down for good.

On the evening of November 24, 1914, Roelofs telephoned the mine attorney, Hildreth Frost, excitedly demanding that he come at once to Cripple Creek from Colorado Springs on a matter of the greatest urgency. Frost arrived the next day, and, in company with Ed de la Vergne, was enlisted by Roelofs to go down the Cresson to view something "unmentionable." They rode the cage to the twelfth level and then walked a half mile to a lateral drift whose course was blocked by newly installed and securely guarded iron doors. Armed guards admitted them, and Roelofs invited his companions to enter a natural cavity at the end of the drift, a cavity evidently disclosed by the last drift round. Lighting a magnesium flare, Roelofs illuminated a room-sized geode whose walls were literally encrusted with crystals of calaverite, sylvanite, and native gold. They estimated in-sight values at $100,000, which proved a bit on the conservative side—after removal of the crystals and milling the quartz lining of the geode, the gross cleanup amounted to $1.2 million.[33]

The Cresson's treasure vault was a pot of gold at the end of a rainbow of dreams, but neither of the chief dreamers had lived to see its wealth discovered. Stratton died a multimillionaire in 1902, leaving an estate that was to be fought over savagely by a host of claimants ranging from an avaricious legislature to women who proclaimed common-law marriage and children, bastardy. Bob Womack lived seven years longer, dying in the same poverty in which he had first entered the district and attracting no estate litigation at all. Coincidentally, both men perished of the same distemper, cirrhosis of the liver, probably induced by the drink. It would appear therefore that if wealth does not bring ease of mind neither does poverty; it merely takes the poor man appreciably longer to get up the price for the last bottle.

Among the landforms favorable to mineralization are major escarpments whose high, planar faces indicate the presence of very long shear planes extending to great depth in the earth's crust. Although the immediate base of such escarpments is usually without values, there may be mineral lodes found in the region of secondary faulting, a score or more miles distant in the low country. One such district is the western margin of the Great Basin, from most of whose mining camps the snow-covered peaks of the Sierra Nevada escarpment can be observed clearly. With the discovery of the Comstock Lode in this zone in 1859 prospectors logically should have begun working south along the edge of the Great Basin. In fact, however, they tended to congregate in the High Sierra gold outcrops at Mono Lake, Lundy, and Bodie. This oversight constitutes little reflection on the desert rats, who were not as a rule given to making broad generalizations. The engineers, however, ought to have looked again at their maps and drawn a speculative thumb down the western edge of the Great Basin.

For forty years after the Comstock strike, southwestern Nevada remained relatively unprospected, or, if examined, nothing of great value was reported there. This cold, windy, and desert basin is, not to put too fine a point on it, terrain in which even the omnivorous Digger Indians frequently starved. With the collapse of the Comstock in the 1880's many jobless miners, including James L. Butler, drifted into cattle ranching in the region. Since a hundred acres of such pasturage will scarcely fatten one longhorn steer, Butler necessarily made the acquaintance of a wide scope of the land while overseeing his herd. Travel in this region must proceed from water to water, and Butler knew of the springs called Tonopah by the aborigines, which lay at the base of the San Antonio Hills. Since springs emerge from geological discontinuities and such discontinuities are also favorable to the deposit of mineralization, one thing led to another. John Hays Hammond summarized it by writing:

> One day as I emerged from the shaft of the Tonopah, I found James L. Butler discoursing learnedly to a party of enthralled eastern tourists on the geological phenomenon which led him to the discovery of the mine. When he

caught sight of me he stopped abruptly. While he was still Lazy Jim Butler, rancher and prospector, he had told me quite a different tale.

I went on my way smiling. Butler had gone back on his burros in a most unsportsmanlike manner. The burro certainly has a "nose for a mine" even though he may not have horse sense; he is a worthy precursor of the modern geophysicist, as in the case of Kellogg's burro of Coeur d'Alene fame.

According to Butler's early version, in the spring [May 19] of 1900 he and his wife set out from Belmont, Nevada, for a near-by district called the Southern Klondike. They camped one evening in a desolate spot at the foot of a hill. The next morning, finding that the pack burros had strayed, Butler set out uphill to find them. In order to speed them on their way back, he picked up a rock and was about to hurl it in their direction when he noticed that it was mineralized quartz.

Mrs. Butler also claimed to be the finder. She said that while she was sitting on a rock pile waiting for her husband to find the burros, she saw a piece of loose rock that glinted with yellow specks. Gallantry impels me to accept her version.[34]

After a number of delays and false starts Butler took in as associates an attorney, Tasker L. Oddie, and a teacher-assayer, Walter Gayhart. Assays indicated outcrop samples going $206 a ton in silver, with traces of gold. The partners located the best-looking prospects and proved them up by location work which hoisted manganese-silver ore for which the smelters paid $500 a wagonload.[35] Butler, Oddie, and Gayhart deliberately trumpeted the news on the correct theory that a mining stampede would create a big camp and that the camp would attract railroad service. The railroad would reduce considerably the unconscionable cost to them of freighting in supplies and freighting out their ore. Tonopah forthwith blossomed like the rose, although, in view of the perpetual shortage of water, its aroma may have differed considerably.

The Tonopah silver strike was the best news Nevada had enjoyed for twenty years. Boomers flocked in and, after locating every foot or fraction of a foot in the immediate vicinity, began to fan out in all directions in obedience to the theory that when one major deposit has been found others may well be not far away. Indeed, in historical fact it is seldom enough that the first real strike proves in the long run to have been also the only, or even the biggest, strike of the district. The prospectors were inclined to look for black manganese silver, however, and so paid little attention to a wide, shallow valley, one long day's march south of the camp. There was, to be sure, a great deal of silicification around and within it—for that matter, a low conical peak centered therein, Mount Columbia, glimmered almost white—but the silica appeared to be of secondary origins and hence uninteresting. The surrounding rimrock was mostly late basalt flows. The valley floor itself was covered with a discouraging amount of unpromising alluvial debris. Some narrow linear depressions following the horizontal contour at the western foot of Mount Columbia contained only a chalky, lightweight substance. Nowhere was there the dead-black stain of the manganese sign for which they sought.

Much later it was concluded that the valley was underlaid by a gently upcurved sandwich of three igneous strata. The upper stratum was andesite, a rapidly cooled microcrystalline rock having about 50 per cent quartz. It was bedded on dacite, a variety of granite. The dacite in turn lay over latite, a feldspathic flow. The andesite cropped out east of Mount Columbia, the dacite emerged at its western foot, and the latite did not appear at all. In any event, none of the three are in themselves regarded as red-hot prospecting country. For the width of two feet or so, however, the outcropping face of the dacite stratum had been infiltrated from below by hot, alkaline waters which had metamorphosed the affected band to alunite. Alunite is a complex potassium-aluminum sulphate, considerably softer than the dacite country rock and hence eroded more rapidly, accounting for the narrow transverse depressions. When presented with a handful of this material, even the most wily prospector will toss it aside, following the act with a swift stroke of the palm across a trouser leg.[36] This particular alunite contained no exotic crystals, had no interesting stain, and exhibited no mass greater than one might expect from a dry, friable, chalky earth. It did carry traces of ocher, a brownish-yellow iron oxide, but ocher commands no interest when in such small quantities.[37]

In 1902 two Tonopah pilgrims, Harry Stimler and Billy Marsh, wheedled a grubstake from James L. Butler for the announced purpose of having a closer look at the valley around Mount Columbia. Why they hit upon this region and what arguments they used on—or against—Butler are conjectural, but they may be perhaps deduced. Certainly the landforms were not attractive, no blossom indicative of

mineralization was visible, nor was the abundant silicification promising. Neither they nor anyone else knew or cared about alunite.
Therefore it seems most likely that they had found float-gold colors at
grass roots near the base of Mount Columbia. To carry speculation
farther, it could well be that Butler questioned them closely about the
float, dragging out the reluctant admission that it was fine-grained,
sparse, and widely distributed. This normally suggested that the lode
of origin might be miles distant, found only with great difficulty, and
not worth a noseful of wet ashes when found. On balance it would
seem that Butler did not really expect any return on his investment
and therefore extended the grubstake more out of sympathy for fellow
prospectors than on the hope of increasing his already considerable
wealth. It is notorious that the West has many such areas containing
gold colors which can be traced nowhere in particular and which are
uneconomical even to placer.[38]

Stimmler and Marsh went to their valley, began operations at the
northwestern sector of Mount Columbia, and at once ran into difficulties. They followed the colors upstream, crossed the alunite trench,
sampled the unaltered dacite of its hanging wall, and found it to be
without values. Backing down the slope, they probably tried a bit more
to the north, advanced upward again, and again were balked at the
dacite. The colors appeared to stop at the shallow declivity, but neither
man conceived that alunite might be the gangue mineral of a lode. At
length Stimmler and Marsh gave up and moved some distance northwest in the valley. They picked up colors again and this time traced
them to a gold-quartz outcrop of modest size and tenor which they
located as the Sandstorm. Perhaps assuring themselves that this outcrop was somehow responsible for all the float color in their Grandpa
District, they rested content.

The moderate Grandpa excitement drew other prospectors down
from Tonopah. They dug and panned around for nearly a year to no
avail. Most of them then retired from the scene, but two persistent
partners, Al Myers and Bob Hart, continued to hang on. If even the
dust of the trail to Tonopah panned a long tail of fine-grained gold
colors,[39] bonanza, they reasoned, must not be far away. In a sense the
entire slope on the west and south of Mount Columbia was a field of

Mount Columbia, Goldfield, Nevada. The Myers-Hart discovery was made at the left base of the peak.

Cripple Creek, Colorado. The granite rim of the crater is visible, exposed by erosion of the breccia on the right.

A typical "rusty" outcrop near Beatty, Nevada. This is what prospectors sought, and many abandoned prospect pits are nearby.

gold. Logic insisted that the float had to originate somewhere up the gentle gradient. Perhaps in mere desperation, or perhaps as part of a program of systematic canvassing of every inch of the ground, in May, 1903, they panned a shovelful of the miserable alunite, recovering what at first appeared to be only the included ocher. Then came a great light into their minds. Ocher is entirely too low in mass to be separated from alunite by rough-and-ready panning. What reposed in the drag might have the color of ocher—might even include some adhering ocher—but was in reality something entirely different and much more massive. When examined under a magnifying glass, the particles showed the heart-warming yellow shimmer of gold so finely divided that it looked brown to casual inspection.[40] The alunite band was a lode, and the values it carried were almost frightening in their enormously high tenor. Myers and Hart immediately staked three locations on the trench at the west side of Mount Columbia, naming their association claim the Combination.

Myers and Hart publicized their strike for the same reason as had Butler at Tonopah, and the big Goldfield excitement was on.[41] After some need to be convinced that the alunite was actually what they were after, the prospectors traced it around the base of Mount Columbia on the southwest, staking every foot of the way. They skimmed off the surface ore by open-cast excavation and then let in the promoters and miners to take it from there. Deep development would eventually unravel the riddle of the Goldfield bonanza. The alkaline hydrothermals, carrying much gold, had originated at great depth and penetrated through the top of a deep shale anticline. Since the shale was not very receptive ground, the mineralized waters pressed up to follow briefly the tight contact between the shale and the latite stratum. Rising through the latite and then following the latite-dacite contact a short distance, they entered the center of the upcurving dacite stratum, which they followed to the surface, emerging in the center of the outcrop.[42] The lode was not so much a continuous sheet, however, as a series of shallowly rising, curved separate ribs. The miners eventually traced each rib down to the top of the shale, a vertical distance of about 920 feet, at which point the values thinned out.[43] The basement rock below the shale is not known, but no one would be greatly surprised if it turned out to be the top of a large magmatic intrusive stock whose rise was associated with the secondary faulting east of the Sierra escarpment.

Goldfield was the last big boom camp of the West, although the doubly encouraged prospectors soon located moderately substantial deposits farther south at Bullfrog, Searchlight, and Chloride (Arizona). Two promoters, George Nixon and George Wingfield, managed to gain possession of the four best properties in the camp, paying over $6 million to merge them into the Goldfield Consolidated Mines Company. That this money was well invested can be inferred from the fact that one railroad carload of raw ore from their Mohawk No. Two was bought by a San Francisco smelter for $574,958.39.[44] An eight-inch alunite ledge in the same development assayed $250,000 a ton.[45] This was the highest of high grade, but good mine-run ore at depth assayed 541 ounces, or 2 per cent of its weight, in gold to the ton.[46] Such high values compressed into so little space was irresistible to the miners, who high-graded so much of the values that the camp was a thieves' kitchen beyond compare. Since the rule of thumb is that under these conditions the miners will steal between 20 and 40 per cent of the values, the official statistics on Goldfield production ought to be increased by one-third to get some idea of the real values hoisted.

The Tonopah-Goldfield boom gave Nevada such prosperity as it had not enjoyed since the Comstock collapsed, nor would enjoy again until the enterprising Benjamin "Bugsy" Siegel conceived of stoping pay dirt in yet another fashion. From first to last Goldfield officially produced $100 million in gold, to which the above-mentioned 30 per cent really ought to be added. Its older twin, Tonopah, lasted a bit longer, hoisting approximately $150 million in silver. Their combined production was therefore $280 million, or two-thirds of that grossed by the Comstock during its flush times. On the other hand, Goldfield and Tonopah unquestionably paid far more dividends, thanks to their comparatively shallow depth and to vastly improved mining, milling, and transportation methods.[47] Goldfield also made prospecting history in another respect as well, nailing down the point that gold is where you find it and that no sort of rock should be despised until it has been thoroughly investigated, be it jewelry quartz or only what appears to be an inferior sort of tailor's chalk.

CHAPTER VII

The Cornish Pump

THE golden age of steam in the United States is usually equated with four prime movers, each of which was a singularly graceful expression of its function and materials. One was the diamond-stack 4-4-0 "American" locomotive, a high-pressure, direct-action engine which by 1860 had approached four hundred horsepower in each of the estimated eight thousand individual units.[1] Another was the shallow-draft, inland-waters, side-wheel steamboat, also propelled by high-pressure and direct drive and of somewhat greater horsepower, but present in fewer numbers. The third was the Hudson River–type side-wheeler, with low-pressure condensing engines, walking-beam power transmission, and blue-water hull lines. The fourth was the stationary mill engine, either high or low pressure and direct action, equipped with a flywheel for power takeoff.[2] All four were much in the public eye, but there a few to recall a very significant fifth power source. This was the Cornish beam engine, a clumsy, almost dinosaurian contrivance, wheezing away unremarked behind the mine dumps. It was not only the common evolutionary ancestor of its racy descendants but came near to outliving them as well. It gave birth to the Iron Age, supplied it with materials, and died out only when the age itself was about to pass away. Like the cockroach, it was very humble but very enduring.

Steam engineering is almost biological in its Darwinian struggle for survival, with sudden quantum jumps of mutation and technical blind alleys, but in broad outline it consisted of five steps which led from its practical inception to the present state of the art epitomized by the ultrahigh-pressure condensing turbine which today supplies about 65 per cent of the world's power. Disregarding the theoreticians, commencing with Hero of Alexandria and extending to the Frenchman Denis Papin, the practical steam engine was peculiarly Celto-British in conception, a fact which seems to have escaped all but historians of technology and poets.[3]

The need for greater power than that which could be generated by water or wind was becoming increasingly urgent throughout eighteenth-century Europe, but was particularly acute in the coal measures of Wales and in the tin and copper mines of Cornwall. In both districts the deepening mines had begun to make water in such quantities that no available countermeasure could keep them economically productive. England's growing commercial prosperity hinged upon assured delivery of brass and coal. Fortunes awaited those who could produce them; only groundwater stood in their way. As in the development of nuclear power, the necessity was so overwhelming that any objections founded on development cost or on the existing state of the art were brushed aside. Money was no object, and, if the state of the art was inadequate, it must be advanced to the point where it *was* adequate. When men speak in this fashion, a thing is as good as done.

The first practical step was taken by the Cornish military engineer Thomas Savery (1650?–1715), who designed a water-raising "fire engine," which was basically a vacuum-and-force pump. The steam raised in an external boiler was admitted past a throttle to a closed receiver vessel having an uptake suction pipe and a discharge pressure pipe, both equipped with automatic nonreturn check valves. When the vessel was filled with steam, the throttle was closed, and cooling water was then induced to flow down over the vessel's exterior surface. This cooling condensed the steam, creating a vacuum which drew sump water upward in the uptake pipe and filled the receiving vessel (the nonreturn valve checked any downward flow thereafter). The throttle was reopened, and the renewed steam pressure drove the vessel's water contents up the discharge pipe (whose nonreturn valve likewise checked any reverse flow). The vessel was now empty of water but full of steam, and the next cycle began, which could be repeated four times a minute.[4]

Savery's fire engine was subject to a number of inherent limitations. The slow cycling greatly limited its capacity, making it useful only to draw domestic water supplies. In application to mine work, the firebox and boiler would have to be located at the surface to permit smoke removal. And since the over-all effective lift was less than fifty feet, a

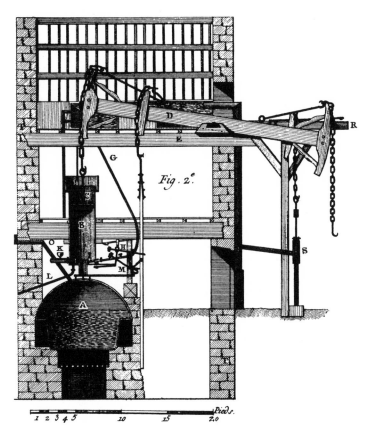

The Newcomen engine, about 1705. *A*, "haystack" steam boiler; *B*, cylinder; *C*, piston; *D*, beam; *E*, beam trunnion supporting truss; *G*, cylinder cooling spray water pipe, fed by overhead reservoir; *H*, spray water valve, actuated by linkage rod from beam; *K*, lubrication oil cup; *L*, cylinder condensate drain line; *M*, steam inlet valve; *R*, segment-of-arc guide for pitwork connecting chain; *S*, lift pump to fill cooling water reservoir. From Denis Diderot, *L'Encyclopédie, ou dictionnaire raisonné des sciences, des arts et des métiers* (1751).

137

number of units working in series could therefore not be spaced down a deep shaft, even though their capacity was scaled up. From the criterion of practical effectiveness Savery's engine was worse than bailing by a horse whim or employing a windmill to operate a conventional lift pump. His real contribution, however, lay in demonstrating that useful work could be accomplished by steam, thus opening the door to further experimentation with an entirely new source of power.

A contemporary of Savery's was Thomas Newcomen (1663–1729), of Devon, who was pursuing the same objective and who was forced by legal considerations into partnership with Savery. Newcomen observed that much was to be gained by separating the power generation from the work-performing function, employing for the latter ordinary lift-pump pitwork which was already in an effective state of development. He transformed Savery's receiver vessel into a true cylinder and introduced into it a moving piston, perhaps borrowing the idea from its originator, Papin. Newcomen sensibly retained the separate boiler, a late idea of his partner. The matter of the piston aside, it might be observed that Newcomen's ideational contribution, whether he realized it or not, was in recognizing that radically different operations were best performed by specialized machinery or, to put it the other way, that an all-purpose system performed none of its functions very well.

Given conventional pump pitwork, Newcomen's "atmospheric" engine constituted a vast improvement in effectiveness upon Savery's vacuum-force pump. As the cycle began, the throttle valve admitted boiler steam into the bottom of the cylinder. Steam pressure was only atmospheric (one to two pounds per square inch, relative); its only effect was to break the vacuum that was holding the piston at the bottom of the downstroke, thus permitting the weight of the pump pitwork to draw the piston upward. (To this extent the steam performed no work—exerted no positive pressure against the bottom surface of the piston. For that matter, the admission of air at this moment to the cylinder at ambient pressure and temperature would have produced the same releasing effect.) At the top of the stroke a spray of water was jetted into the steam-filled portion of the cylinder below the piston, condensing the steam to produce a vacuum. The pressure of the atmosphere (fifteen psi, relative) freely exerted against the upper side of the piston at once drove the piston back down to the bottom of

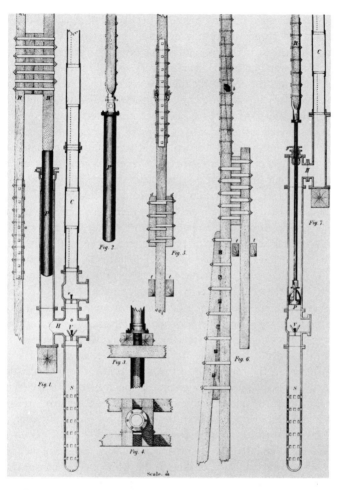

Cornish pitwork pumps. At the extreme left is a section of a Cornish force pump in its "aitch" (H) casing. At the extreme right is a lift pump in its casing. Both pumps terminate in "eggs," or pierced intakes. Figs. 3 and 6 indicate the complex timber splicing of such pitwork. From Clarence W. King, *United States Geological Exploration of the Fortieth Parallel.*

139

the cylinder, generating the engine's effective work. A connecting rod and first-order walking-beam lever linked the piston to the pump pitwork, so that, as the piston end of the beam was pushed down, the beam's other end raised the heavy pitwork against the force of gravity, thus "cocking" the engine for the next cycle.[5]

Newcomen's atmospheric engine was itself not without faults, but these were flaws of efficiency, not practical effectiveness. A Newcomen engine of sufficient size could operate pitwork extending downward for several hundred feet and could generate the power to raise large quantities of water at each cycle. Precise information is unavailable, but it has been surmised that these engines' nature and utility were widely recognized within fifteen years and that "several scores" of them persisted in employment for fifty years.[6] It is probable that the complete adaptation of mine pump pitwork to steam power was likewise accomplished during this period, changing very little thereafter until its complete displacement about 1885 by totally different devices.

The third quantum jump in steam engineering was made by James Watt (1736–1819). His career is strikingly similar to that of Rudolph Diesel in that both men approached their ends from theoretical rather than practical considerations, both taking a remarkably wasteful power source and revamping it so fundamentally that within their own lifetimes they reaped substantial financial and other rewards. In Watt's case, however, the canny Scot actually performed two separate feats: he first made the Newcomen engine so much more efficient that it remained virtually unchanged until 1900, and then he went on to develop a revolutionary second steam engine of a much different and higher order—this last rather despite himself.

Watt discerned that the chief fault in the Newcomen engine lay in the successive heating and cooling of the cylinder. This consumed so much additional energy (which was not expressed as useful work) that the system was highly wasteful of fuel. A comparison could be a hypothetical gasoline engine, half of whose power was diverted to providing electricity for the ignition system. Such waste was acceptable in the Welsh collieries, where unsalable culm and slack were available for stoking, but it was financially ruinous to the Cornish mine adventurers (investors), since coal not only had to be imported but had to bear an idiotic revenue duty as well. Watt continued the Newcomen principle of separating dissimilar functions by creating the necessary vacuum in a separate condensing vessel, which was cooled by a continuous water flow. At the top of the piston stroke a cylinder exhaust valve was opened in the vacuum line leading to this condenser, and then the valve was closed at the bottom of the stroke. This improvement enabled the steam cylinder to be maintained at a constant temperature, with consequent great saving in fuel: the steam cylinder was always hot, the condenser always cool. It also eliminated "dead time" necessary for the previous cylinder-wall heating and cooling and hence speeded the cycle and much increased the hourly work capacity.

Apart from the addition of the condenser and a few auxiliary parts, there was no other change in the engine, but the firm of Boulton and Watt realized a handsome profit by charging the owners of the Newcomens they modified a royalty of one-third the subsequent coal saving. To ensure that there was no cheating, Watt calculated the engine performance before and after modification with a series of fuel-consumption tests prorated to "duty" (effective work performed per cycle). Knowing the fuel saving per cycle, all he had to do was know the number of cycles or strokes the engine made in, say, a month. Watt invented an ingenious counter rather like a pedometer, which was attached to the walking beam and was securely locked to prevent tampering. Once a month his agent came around and read the meter, after which the bill was made up and submitted.[7]

With the money obtained from this tour de force, Watt went on to make the third quantum jump in steam engineering by inventing the double-action engine. Its principle was the application of a Newcomen atmospheric impulse in alternation to each side of the same piston. This doubled the effective energy output per cycle and permitted still higher speeds. It was also capable of producing a useful rotary motion through (ultimately) a crank and flywheel transmission and was the progenitor of all "low-pressure" (condensing) double-action steam engines.

The fourth step was taken by the railroad locomotive designer Richard Trevithick, who had to discard the condenser (no cooling water was available) and therefore operate the double-action cylinder by force of direct steam pressure, thus creating the "high-pressure" direct-action engine whose speed and capabilities were limited only by the strength of the boiler.[8] There matters rested until, as the fifth and final step, Sir Charles A. Parsons, about 1900, brought the wheel full circle by devising the rotary steam reaction turbine foreshadowed by Hero of Alexandria. With its one moving part, ultrahigh pressure, and condenser vacuum, the turbine has since been improved only in detail.

The Cornish mine beam engine, however, remained to 1810 no more than a Watt-modified Newcomen Model 1775, which is to say single-action, vacuum-condensing on the power stroke, with the return-stroke impulse provided by the weight of the pump pitwork. The expiration of the Watt patents thereafter permitted engine wrights and their patrons leeway to indulge their fancies with any combination or permutation of the two Watt systems they thought desirable. Presently, therefore, double-action direct pressure gradually came to prevail, though in Cornish installations the condenser was usually retained, since its presence gave an additional fifteen psi effective pressure differential without imposing its strain on the very dubious boilers available.

These beam engines were large even in Newcomen's time, and they grew gargantuan before the end of the nineteenth century. The deeper the mine sump the relatively more fragile became the pump pitwork and hence the fewer the cycles per minute that could be attempted. Therefore, to increase pumping capacity and height of lift, deliverable power per stroke had to be increased. This was done by enlarging the steam cylinder or by compounding a number of cylinders.[9] Cornish adventurers preferred simplicity and greater size, particularly since the accessibility of the shire to the sea meant that really huge cylinders and beams could be transported with relative ease to the site of installation. Accordingly, cylinders of 90-inch diameter were not uncommon, while one mammoth cylinder of 144 inches was cast in 1843 for a Dutch land-reclamation pumping station. Such monstrous trusses and castings could not be employed on the American frontier, where trails were abominable and freighting costs alarming. Thus some fundamental changes would have to be made by Americans on this account alone before they could adopt Cornish pumping.

The conservative Cornish preferred simplicity in the engine's power transmission as well. Excepting some late convolutions the engine house was built of stone adjacent to the mine pump shaft.[10] The steam cylinder was erected in the house, the beam trunnions bedded in the heavily reinforced front bob wall, and the nose end of the beam extended out a large second-story window (the bob opening) so that it terminated directly over the pump shaft with the upper end of the pump rod coupled directly thereto by a massive pivot. Laminated wooden beams gave way to cast iron and cast iron to built-up steel trusses weighing over twenty tons, but had James Watt been introduced to a Cornish engine house of 1920, he would have been comfortably at home with everything but the electric lighting (should there be any) and the substitution of pivots for the old connecting chains at the ends of the engine beam. Watt had feared pivots and cranks (as well as suits for patent infringement), thinking that, since a power stroke was subject to some variation in travel, rigidly limiting the travel of the other end of the connecting rod might well produce a smash. For this reason he retained Newcomen's system of coupling the pitwork to the beam with chains, which groaned in an unearthly manner as they worked over their great segment-of-arc guides. For his "rotative" mill engines he devised an ingenious but overly complicated planetary gear linkage to the flywheel, again both to avoid patent trouble and to permit "play" if the stroke varied. Experience soon showed that his fears were unfounded and that simple pivots and cranks would do very well if forged sufficiently strong.[11]

Power transmission set in with a vengeance in the pump pitwork itself. Speaking broadly, the backbone of the pumping system was a massive wooden pump rod, measuring as much as sixteen inches by sixteen inches in squared cross section. It extended from the nose of the engine beam down nearly to the sump and was of course built up by splicing together timber sections, reinforcing the points of extra weakness with all manner of iron straps, bolts, and clamps. When in action the rod had a vertical movement of as much as twelve feet.[12] Since it was inherently flexible and fragile in consideration of its length-to-cross-section ratio, it was held in alignment by well-lubricated timber guides spaced down the pump shaft at frequent intervals. Experience demonstrated the desirability of fastening balks of timber called "wings" crosswise to the rod at intervals so that if the rod broke the wings would catch in the shaft timbering and so prevent the rod from collapsing like a handful of jackstraws into the sump. A large, man-powered hoisting whim or capstan was installed near the shaft, as were tripodal shears, so that the rod could be disconnected from the engine beam and hoisted by hand for repair or maintenance.

The rod terminated in the mine sump with a simple lift pump, the connecting rod and casing of which could be readily extended downward by bolting on standard iron intermediate sections when required by deeper sinking in the sump. The water was drawn in through a pierced intake strainer (an "egg" in Cornwall, a "niggerhead" in the American West) and was raised atop the piston to the limit of the effective strength of its materials. The intake valve and piston check valve of this "drawing-lift" stage were two simple flaps, or "clacks," made reasonably watertight on their seats with a leather facing. Renewing the clack leathers was frequently necessary, and some weird experiments were made in an effort to find the most acid- and wear-resistant leathers: it is said that hippopotamus hide enjoyed a brief vogue for this purpose.

Once the water column had been raised to the limit possible by the drawing lift, intermediate stage pumps could be bolted to the rod, each drawing from a reservoir supply tank fed by the rising main leading up from its next-lower neighbor; the beauty of the system was that, as the mine went deeper, the pitwork could likewise be extended downward without need for a complete revamping until the breaking strain on the rod itself was approached or the engine became clearly inadequate. These intermediate stages were not lift pumps but plunger-force pumps. As the rod descended under the pull of gravity, the plunger bolted to it slid down into the pump casing, which was shaped vaguely like the letter H (hence, its name). The force exerted by the plunger pushed the water contents of the H up the rising main. As the rod paused, the valve clacks within the H reversed position, checking any backflow and permitting more water to be drawn into the plunger body by suction from the reservoir tank when the rod began to rise.

In deep mines the combined weight of rod, plungers, and water column aggregated astonishing tonnages. The Tresavian copper mine in Cornwall had a deep, though not exceptionally so, pump shaft whose over-all depth was about fifteen hundred feet. The rod was equipped with nine water lift stages, presumably including the drawing lift in the sump. The weight of the rod and plungers was 57.5 tons, and that of the water column was a bit more than 36 tons. It must have therefore discharged at the surface 4 tons of water a cycle, and, although its cyclical rate is not given, it could not have been more than seven or at most eight strokes a minute. This work was accomplished by an 86-inch-piston-diameter steam cylinder—large, but not uncommonly so.[13]

Such aggregate tonnages were far more than were required to draw up the engine piston. In fact, they would have seriously unbalanced the machinery without some alleviating factor. For that matter, engine beams frequently failed at the trunnions or the gudgeons (pivot bearings) as the result of comparatively slight and hence unnoticeable imbalance. The solution adopted was to attach a system of variable counterweights to the rod and certain other parts of the pitwork. These consisted of massive levers pivoted upon a fulcrum, whose one end was connected to the pitwork by another pivot. The free end had a heavy open box into which scrap iron could be loaded in whatever quantity was thought to achieve the desired degree of static balance in the system—in the Tresavian pump the counterbalance tonnage was 140. Such balancing was never the product of formal calculation, but was done by the pumpman, who looked, listened, experimented, and finally came up with what he felt was the right counterload. That such intuition was often incorrect was demonstrated by the frequency of beam and rod failures, most of which could be attributed to dynamic imbalance.

The working of the Cornish pump appears unnatural to minds conditioned to the modern practice of accomplishing work by the direct and positive application of energy to the desired end. The system appeared at first glance to do everything backward and by the relief, rather than by the application, of force. The Watt-modified Newcomen engine admitted steam only to break the vacuum that was holding it motionless. The release of the piston permitted the rod and plungers to ease downward in the pump shaft under the influence of gravity. Nonetheless, this was the work phase of the cycle, for the descending pitwork drove the column of water up the rising mains and at the same time raised the weighted ends of the counterbalance levers. When the engine exhaust valve was opened, the vacuum in the condenser drew down the piston and pulled up the pump rod (plus the drawing lift piston and its quota of water), assisted somewhat by the dropping of the counterweights. Yet all this had to be so, for while the rod possessed some degree of tensile strength, it lacked for all practical purposes any rigidity or strength in compression. It is a pity that working models of Cornish pumps and pitwork are so rare in the United States, for such a model would draw more attention than a dog fight and outdo a Calder mobile for its apparently aimless complexity of motion.

That such a marvelous mechanical monstrosity persisted in mining work and demolished all proposed competition the world over for 150 years was the result of three advantages, two of which were important, but the third of which was decisive. First, the system was relatively simple in principle (if not application) and could be erected, operated, and repaired by analphabetic Cornishmen who would have migrated to hell on the prospect of a job in mines in which it was employed. Second, it was remarkably economical in fuel consumption, a matter of great moment in most nonferrous mine developments during the nineteenth century. Third, and critically, the Cornish pump was the only system then capable of transmitting effective amounts of power down the pump shaft of a deep mine to the points at which the power had to be applied in order to force the water upward. There was no other method within the state of the art of so transmitting useful energy to the extent and in the amount required. As the proverbial traveling salesman remarked upon joining the admittedly dishonest poker game, it was, in the last analysis, the only game in town.[14]

The tale of complexity is by no means finished at this point, since it was seldom enough that a pump shaft could economically be sunk in true vertical plumb to the better accommodation of the delicate pump rods. More often than not, for various reasons, such shafts had to be inclined, or turn a number of angles, before terminating in the sump. It is not possible to make a sixteen-by-sixteen-inch timber rod slither around a corner like a snake after a rat, desirable as it may be. The "old men" of Cornwall solved this vexing problem by wide and ingenious employment of bell cranks, which they called "bobs," as they did every moving truss, whether engine beams or counterweight levers. A bell crank is a rigid triangle, pivoted either at one corner or in the center of its hypotenuse to a fulcrum. Under proper conditions a push or pull exerted on a second corner will be transmitted from the third corner, at any angle up to 180 degrees from the original axis of motion, through movement of the entire truss. The bell crank is still widely employed in keyboard devices such as manual typewriters, where the same problem of making a mechanical impulse turn a corner without appreciable loss of energy is encountered. Another application is seen in railroad marshalling yards, where the turnout switches are actuated by hand levers in a central tower, the impulses pulling or pushing very long rods supported on rollers, the rods terminating in a horizontal bell crank opposite the switch points.

These bell-crank bobs could almost literally take a rod down a spiral shaft, should such a work have existed. Their moving corners served as convenient points on which to hang yet more counterweight boxes. They could transmit power horizontally by "flat rods" at grass to drive hoists, or "winding whims," once the "old men" finally brought themselves to employ such devices. Correctly linked, a pair of bobs could drive parallel pump rods in opposing directions of travel within the shaft, giving rise to their employment as "man engines." Invented in the Saxon mines, a man engine consisted of two pump rods in opposing motion, each equipped with hand grips and foot boards spaced down at twice the distance of movement. With some practice a miner could ascend by riding up one rod to its limit of travel. As it and its fellow paused for the couple of seconds allowed to permit the pump clacks to reverse and seat themselves, he stepped over to the other rod, which was now at the bottom of its travel. It would rise, carrying him upward. Zigzagging back and forth in this manner, he would be carried steadily up the pump shaft without personal exertion.

Mines with single pump rods achieved the same end by having fixed platforms built in the shaft timbering next to the rod. A miner stepped to such a platform at the end of an upstroke, waited out the downstroke, and then resumed his journey at the next ascension of the rod. While a modern safety engineer would regard this procedure as deplorably dangerous, the Cornishmen far preferred man engines to the labor of climbing hundreds of feet of ladder at the end of the day. Since wire rope, which made raising men by steam hoists reasonably safe, was developed soon after the opening of the American mineral frontier, man engines were very seldom used in the western mines of the United States.

The Cornish mine adventurers were not fond of "rotative" engines, as they called them, although they found rotary-power takeoffs indispensable for running stamp mills, concentrating machinery, and winding whims. Railroad designers needed flywheels no more than did the Cornish mine pumps: the locomotive itself accumulated and conserved momentum in its forward travel, thus smoothing out the bumpy effect of the piston impulses. Watt discovered, however, that his mill engines demanded flywheels to ensure a smooth delivery of power. Around 1804 two different basic flywheel arrangements were coming forward. One, the more familiar, made the flywheel an integral part of the crank power transmission which drove the line shaft or belt pulley. The other system placed the flywheel on the cylinder side of the beam, independent, to all appearances, of the power train, and at first glance such "balance wheels" appeared to have no function but to delight the eye with their rotation.[15]

The flywheel was destined to become critical in the American West, however disregarded it had been in Cornwall. The reason was economic rather than technical. It has already been emphasized that beam engines directly coupled to the pump rod had to enlarge in proportion to increasing shaft depth and pitwork load. As matters stood, there seemed no way to avoid this problem, nor did the Cornishmen, with their enjoyment of cheap water transportation, have any reason to do so. In the American West, however, overland freighting charges were one of the most important costs of mining. By way of illustration it may be noted that in 1859 the Ophir Mine of the Comstock Lode was shipping silver concentrates assaying three thousand dollars to the ton but was paying close to one-third of this sum for wagon and boat transportation to San Francisco.[16] Cost accounting was unheard of, and American silver nabobs were not noted for thrift, but that was a bit much, particularly since the Ophir probably got a favorable rate from teamsters who would otherwise have had to return empty to Sacramento. On the basis of that rate a Cornish beam engine of modest capacity, aggregating fifty tons for cylinder, beam, boiler, and pump ironwork, delivered to Virginia City in 1862 would cost in freightage alone from the wharfside about fifty thousand dollars, whereas a somewhat larger engine could be bought secondhand in Cornwall for a mere six thousand dollars.[17] In a word, it would not do.

For that matter, all the machinery used in the mining West had to be cheap, sturdy, and, especially, lightweight. A mine captain at "'ome" might boast of the massive cylinder and beam of his engine, but if he attempted to carry his theories to Nevada or Colorado, he would instantly find himself among the unemployed. The Blake jaw crushers and California stamp mills avoided cast-iron beds and framework in favor of timber well mortised and secured with bolts.[18] The same principle was true for concentration machinery—and even for railroads, which were narrow-gauged to secure the tremendous economic advantage of reducing mass by the cube root as dimensions decreased arithmetically.[19] For all practical purposes there were but three seaports of entry to the mining West: Sacramento, California, Independence, Missouri, and Guaymas, on the Gulf of California.[20] An ellipse touching these three points enclosed most of the mining frontier, some of whose camps were a thousand miles from navigable water. And until 1869 no camp of any consequence was served by railroad.[21]

The California and Montana placers had no ground-water problem with which primitive "Chinese" chain pumps[22] driven by waterwheels could not cope, but the Comstock Lode presented an entirely different problem as a result of its vast flooding. The pioneer Ophir development struck water at the fifty foot level, and to handle it there was installed the first fifteen-horsepower steam hoist in Nevada, bailed with a self-draining water skip.[23] When this proved inadequate, the Ophir developers next drove a gravity drainway adit to the east in characteristically Cornish fashion. By 1865 this palliative had likewise ceased to be effective, for the levels went below it.[24] Steam pumping was recommended by the manager Philipp Deidesheimer, but the solution achieved in Mexico and Australia of installing secondhand pumps purchased from the failing Cornish mines was not attempted. For that matter, it was stated in 1877 that not one Cornish engine was to be found in the American West.[25] If the Comstock mines were not using Cornish pumps, what were they using?

The paradox is unraveled if it is assumed that the observers were purists who refused to accept as "Cornish" any system which failed to meet the criteria of single-action condensing engines with direct linkage of the beam to the pump rod. On the other hand, there exists a very clear mechanical drawing of the Comstock's Savage Mine pump, which

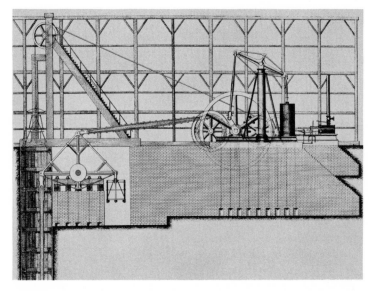

The rotative Cornish engine of the Savage Mine, Comstock Lode, about 1862. The equipment on the extreme right is part of the hoist engine and has no connection with the walking-beam rotative engine. Note the exceptionally massive foundation work required and the counterweighted bob connected to the pitwork rod. From Clarence W. King, *United States Geological Exploration of the Fortieth Parallel.*

was undoubtedly installed about 1862 in the "second-line shafts," where steam pumping was first employed. The cylinder shown is about thirty-four inches in piston diameter, which is far too small for the post-1870 "third-line shafts." The drawing does not even suggest the valve gear, but this was in all probability appropriate to noncondensing direct double action, the kind most familiar to Americans from their railroad and inland-waters high-pressure practice.[26] For that matter, the lack of an adequate (or, indeed, much of any) water supply for Virginia City until 1873 virtually prohibited the use of condensers, whose cold-water requirements are considerable.[27]

Of great positive significance in the drawing is a clearly indicated reduction gearing and a flywheel, the latter mounted on the same shaft as the small driving pinion gear, which is coupled by a crank to the engine beam. With a tooth ratio of about three to eight, the large driven "bull gear" is connected by a horizontal pitman rod to the bell crank above the pump shaft. This crank is properly equipped with a counterweight box, and from its third corner the single pump rod hangs in traditional Cornish fashion. It is often unsafe to generalize from a single example, but since the Comstock mines were given to imitating each other's methods very closely,[28] it is probable that the rest of the big silver producers enjoyed somewhat similar arrangements. Accordingly, the purists can be refuted by the assertion that the Savage pump system was Cornish in every respect but the omission of a condenser and the inclusion of a geared-down "rotative" power transmission.

This being the case, the economic considerations behind this modification at once reveal themselves. The cylinder, boiler, and engine beam could be significantly miniaturized and lightened, as long as a flywheel and reduction gearing were provided. What a "true" Cornish engine accomplished in one power stroke at the expense of enormous deadweight this much smaller and lighter design would do equally well with five power strokes, whose energy was accumulated and smoothly delivered by the flywheel. To be sure, a flywheel was necessarily massive, but it could be cast and transported in segments to be bolted together at the site of use, whereas a cylinder could not, and boiler and beam only with difficulty. The speed of rotation of the flywheel was customarily so slow that the assembly bolts would readily hold against the low centrifugal force exerted. In fact, the most massive single casting other than the cylinder to require transportation as a unit would be the bull gear.[29]

One additional step toward economy in cost and transportation is suggested in a charcoal sketch of the pump transmission at the Gold Hill Mine at Grass Valley, California, by the artist George Mathis.[30] It shows reduction gearing on the Comstock pattern, but in this case the small driving pinion itself shares a common shaft with a large composite wooden belt pulley of the sort commonly used in the West for leather belt drives. The belt sides taper toward each other significantly at the left edge of the sketch, something more than a suggestion that its own drive pulley was smaller again. Here is evidently a double-

151

reduction drive train having an over-all ratio of perhaps ten to one. This would make pumping from moderate depth well within the capabilities of an ordinary horizontal mill engine, which, if purchased secondhand, would be a very inexpensive power source. The caption quotes the famed Melville Attwood (whose assay house at Placerville, California, first reported the silver values of the Comstock Lode) to the effect that the first Cornish pump in California was erected at this mine in 1855 and that the castings[31] were poured at a small foundry which he himself put up near the Gold Hill Mine.

The Cornish engine and pump were subject to various accidents, some of which could be very ugly. When using high pressure, the worst mischance was a boiler explosion, a lamentably frequent occurrence before 1870. Watt had feared and fought high pressure for this very reason, but efficiency and simplicity outweighed his arguments. An explosion usually was initiated when the throttle was closed to stop the engine temporarily, but the fires were continued unchecked in anticipation that it would soon be started up. If the water level in the boiler was too low, the fire's energy was soon diverted from making steam to superheating the existing steam that filled the boiler shell. This produced an insidious rise in internal pressure which checked the vaporization of the remaining water and began to superheat it in turn. Lacking effective pressure gauges, the increase would go unnoticed. Worse yet, it usually checked the inflow (if any) of new feed water by creating feed-line back pressure sufficient to overcome the relatively weak impulse of the crude steam-jet injectors then in use. With but a few minutes of continued firing and buildup of pressure all elements for a disaster would be present.

When the operating engineer reopened his throttle, the slight pressure drop permitted the surface of the superheated water to flash into steam. This instant surge at once ruptured some part of the boiler shell, permitting a yet greater pressure drop, which now flashed all the remaining water to steam—volume for volume, such water held greater potential energy than an equivalent amount of dynamite. One of two equally inconvenient things would then occur. The boiler would burst wide open, releasing a deadly cloud of steam and shrapnel-like iron fragments. Or, if the rupture was confined to one area, the boiler would rise like a rocket and for the identical reason. On Mississippi

River packets such as the ill-omened *Sultana*, where as many as eight such boilers were interconnected by one steam line lacking over-pressure check valves, the entire boiler line would go. An equivalent effect could have been gained by a stick of five-hundred-pound aerial bombs exploding in the fireroom.

Safety valves, fusible plugs, and positive-feed water injectors were invented to prevent such disasters, but in practice each was subject to limitations. The safety valves were conical plugs held down in a seat atop the boiler shell by second-order levers near whose free end heavy weights adjustable in position were hung as with a single-beam balance. Overpressure would lift the plug against the leverage upon it, allowing the steam to blow off before it was dangerously superheated. However, the plug might rust into its seat, or even be soldered into place by a wicked engineer, or the weight itself might be feloniously increased by hanging a heavy wrench on the end of the lever. Fusible lead-alloy plugs swedged into rivet holes below the permissible water-level line would melt and blow out, releasing pressure, if the water level dropped beneath them and so no longer kept them relatively cool. Americans were never as fond of this device as were the British, possibly because a plug blowout meant that the affected boiler had to be taken off the line and cooled down for replacement. In packet-boat days, time was money, and the loss of a day's operation was resented by the owners.

Mechanically actuated positive-feed injection pumps were called "doctors," since they were regarded as a cure-all for boiler difficulties, but to be effective they had to be supervised by an intelligent and conscientious man. Even given an engineer who minded his business, the only way to check boiler pressure and water levels was to listen to the sound of the engine and periodically open a set of small gauge cocks spaced vertically in the boiler from maximum to minimum water levels. If water did not spurt from the lowest gauge cock when it was opened, it was time either to start the "doctor" or to jump overboard, depending on a hasty estimate of the situation.

Accidents in directly connected Cornish beam engines most frequently resulted from structural failure of the engine beam under overload or its equivalent, an unbalanced load. This was particularly true during the period from 1830 to 1870, when beams were made of cast iron. The earlier laminated wooden beams and the later steel

TOMBSTONE PROSPECTOR

The Cornish pump engine at the Grand Central Drainage Shaft, Tombstone, Arizona. The engine has stood idle and in disrepair for some time, indicating the date of the photograph to be after 1890. The engine is of the later, balance-wheel design. Courtesy Rose Tree Inn Book Shop, Tombstone, Arizona.

trusses were sufficiently elastic to give audible and visual notice of trouble, groaning and cracking when abused. Not so the treacherous cast iron, which would snap without any warning. These failures usually occurred either in the gudgeon at the nose (pump end) of the beam or in its center at the trunnions. If the nose failed, the entire beam might lash back into the engine house, coming down on the cylinder and valve gear like a rogue artillery piece in recoil. If the beam failed at the trunnions, the forward half could well fly outward and right down the pump shaft, carrying away the pitwork and timbering as it smashed its way to the sump. Any man in the path of either event was likely to have a very bad time of it.[32]

A perpetual source of trouble in the pitwork itself was dry rot, a concomitant of damp heat. Dry rot can insidiously change an apparently sound timber into a papery sponge. Its ravages can be detected only by frequent, inch-by-inch probing of the timber with a knife point or awl. Should the point sink deep into the wood without appreciable effort, instant and radical surgery is needed. The affected timber and its immediate neighbors must be removed and replaced, for this vile fungus is contagious beyond belief, and there is no way to salvage or patch up wood so infected. The greater part of the pumpman's daily labors consisted of climbing up and down the hundreds of feet of accommodation ladder paralleling the rod in the pump shaft, constantly probing for dry rot and almost as constantly renewing the clack leathers in the H's and the drawing lift.

A second frequent cause of pitwork failure was the pumpman's carelessness in inspecting the water level in the sump. Should the pump continue to work after the water level went too low, air would be drawn into the lifts and rising mains, reducing the load and destroying the balance of the pitwork and causing the engine to race, or "beat." One observer described this phenomenon and offered by inference a comment on careless operational practice in Cornwall.

The greatest and most frequent cause of the engines beating (or as the North countrymen call banging) is, when the water is in fork and the air gets in with the water into the pumps, the engine then wants her due load, raises lighter and quicker, also returns quicker. . . . The damper is of great use in regulating the fire, but never used as the immediate cause of stopping the engine, for that only causes the fire to slacken slowly, whereas the engine when it begins to beat must be stopped in an instant or great damage will ensue.[33]

Such racing could not only damage the engine but work havoc on the pitwork as well. If pump speeds were so deliberate that a pumpman could say of his equipment, " 'er will do 6 [cycles per minute] but it do strain 'er,"[34] it is evident that "she" would be worse than strained when raced out of control.

Whether initiated by dry rot, air in the rising mains, imbalance, or whatever, pump rods frequently snapped. In a true Cornish engine a failure near surface might release so much load as to throw the beam back off the trunnion bearings with results equivalent to a fracture at the nose. In an American rotative engine, the effects might be even more spectacular. American engines used no governor, the flywheel and pitwork serving that purpose effectively enough. But if the rod broke off short, this restraint was at once released. The engine would suddenly race free of load, accelerating the flywheel to an ever-increasing speed of rotation. If the operating engineer could not close the throttle in time, centrifugal force would begin to fracture the flywheel assembly bolts. The wheel would distort and then disintegrate with a roaring smash that would take the roof off the headframe house and fill the air with steam, splinters, and flywheel segments.[35]

The gradual spread of the standard-gauge American railroad network in the West after 1870 produced a somewhat reactionary return to pump-engine designs of the original directly linked variety. Engines installed after that time show dramatic enlargement, the adoption of condensers, and a movement away from reduction gearing. Flywheels were retained but were moved over to the cylinder side of the engine to act as balance wheels. If adequate reserve power was available, the balance wheels might drive belts to take off power for other machinery. Giantism became pronounced, both at the wet mines of Tombstone, Arizona, and in the Comstock Lode, where the "fourth-line" pump shafts were going as deep as twenty-five hundred feet.

It is often assumed that, upon its completion in 1878, the great Sutro Tunnel effectively drained the Comstock mines. That is only partly true. The levels then being stoped were near or at the drainway's depth of seventeen hundred feet when it was opened for business. As long as this was the case, the Sutro Tunnel drainway was effective. In time, however, exploratory levels would be sunk as deep as one thousand feet below the Sutro, encountering floods that had to be lifted that much just to make it possible to reach the tunnel. Between 1878 and 1885 the Comstock mine owners felt it necessary to combine their resources, sink new common pump shafts on the "fourth line" to serve three or four developments, and thus share the enormous costs of the work and of installing the greatest Cornish pumps perhaps ever to be seen in the West.

One such enterprise was the Union master pump shaft financed by the Union, Sierra Nevada and (New) Mexican mines. Its sump was blasted out on the twenty-five-hundred-foot level in 1878, so as to lift the water accumulating there eight hundred feet upward to the Sutro adit. The engine itself was compound-condensing with direct linkage to the rod, but the cylinders were separated, with the sixty-four-inch high-pressure cylinder at one end of the beam and the great one-hundred-inch low-pressure cylinder at the other, working in alternate seesaw fashion. The maximum piston stroke was eight feet three inches. The massive balance flywheel was forty feet in diameter and weighed 110 tons, while the built-up wrought-iron beam was forty-eight feet between gudgeon centers and was supported on a trunnion rod twenty-two inches in diameter.

The pitwork was equally impressive, having an over-all length of twenty-five hundred feet. The counterweight balance bobs were spaced down at four-hundred-foot intervals in the shaft, while the stage pumps were spaced at two-hundred-foot intervals. Thus, when in operation, the pitwork had one drawing lift and but three plunger pumps with their associated reservoirs. The total pitwork tonnage (probably exclusive of counterbalance mass) was approximately five hundred tons. Because of the relatively short eight-hundred-foot lift, the pitwork was capable of ten cycles a minute, a comparatively high speed, and delivered two million gallons to the Sutro Tunnel each twenty-four hours. The fuel consumption in the twelve boilers was thirty-three cords of wood a day, but the operating cost per ton of water raised was a remarkably low nine cents.[36]

This fuel economy must have been of some consolation to the stockholders who paid the assessments required to capitalize this work. The over-all cost of the Union installation probably equaled the six million dollars known to have been spent on the similar Combination shaft and equipment. Lamentably enough, the dividends declared as a result of this, or for that matter any working below the Sutro Tunnel level, were nil. The Comstock Lode had practically been worked out when the Sutro was opened, and these subsequent developments were projected far more on mere speculative hope than by reference to available geological knowledge. Among themselves, the three mines of the Union shaft hoisted after 1878 but four thousand tons of silver ore, assaying from eight to thirty-five dollars a ton. This did not even repay the direct mining costs, meaning that the entire capitalization of the drainage systems was lost. As technical feats the Union and Combination shafts were unsurpassed for their day, but their economic utility was zero. They were magnificent—but that was not mining.

CHAPTER VIII

The Desert Canary

L IVESTOCK of all sorts was essential to the provision of transportation, food, clothing, security, and even companionship to the men who went to the American frontier. With one or two highly qualified indigenous exceptions[1] these mammals and birds were of Eastern Hemisphere extraction. Though the American Indian woman was one of the world's great plant breeders, the Amerind, generally speaking, seemed to have the same blind spot toward the domestication of animals as he had toward the smelting of metals. The Indian's opportunities were limited by the lack of suitable species—the larger fauna in the Americas at the time of Columbian discovery was curiously limited and unbalanced, many major ecological niches going unfilled. Fossil remains suggest that before, say, 50,000 B.C., the Western Hemisphere had a mammalian fauna as varied and colorful as one could wish but that some subsequent nameless catastrophe exterminated perhaps dozens of larger species as diverse as giant sloths and saber-toothed cats. For instance, there was in 1492 no American major predator functionally comparable to the Asiatic tiger or the African lion, although both the puma and the lobo wolf were clearly evolving in that direction. To give another illustration, the huge numbers of bison and passenger pigeons strongly hint at the absence of the population controls which are at work in a mature ecology. It might even be argued that the North American Indian was biologically competing with the grizzly bear for the niche of major omnivore, the Indian not always winning.[2]

The aborigine therefore had few animals about the house. He brought the dog with him on his migration from Asia. In tropical America he tamed (but did not breed) parrots, which, with dogs, served as pets, camp sentries, and acceptable entrées. In the high Andes Mountains the llama and its relatives were pressed into transport service and sheared of their wool. Given more time, the Indian might have fully domesticated the turkey, the mallard duck, the Canada goose, and perhaps even the bighorn sheep and the beaver, but he had available no indigenous species of hoofed animal large and intelligent enough to make into a source of transportation and power comparable to the horse or camel. The profoundly revolutionary effects on the Plains Indian produced by the reintroduction of the horse by the Spaniard indicates his pressing requirements. It is reported that the Indian of the Southern Plains referred to the horse as the "god-dog." Evidently the dog was his only burden bearer and standard of comparison, while "god" seems obviously a desperate attempt by a puzzled interpreter to reconcile two totally different semantic concepts. Was the Indian really saying "burden bearer of inexplicable or divine origin?" In view of his philosophy and of what the horse meant to him economically, that appears reasonably plausible.

The mammalian poverty of the Americas made it necessary for the Europeans to import their own useful (or in some cases deplorably pernicious) fauna. The Spaniards brought the horse, ass, sheep, goat, pig, and cow. They unintentionally gave free passage to rats and mice. Since the traditional organic antidote for *Rattus* and *Mus* is the small predator *Felis domesticus* of Sudanese origin, Pussy appeared on the passenger list. It is reputed that in early Texas times a pregnant lady cat was considered essential to a ranch and accordingly commanded a high price. Her progeny then assumed the duty of protecting the oat bin and providing amusement to the children in return for occasional subsidy and toleration of a nocturnally irksome life-style by the ruling establishment. It has gone unremarked that such casual and unmalicious importation of European-Asiatic fauna by the early settlers would have soon destroyed the precarious ecology of the Americas irrespective of the intentions of those involved. It has been underemphasized that this fauna necessarily included contagious microorganisms to whose onslaughts the Indians, thanks to their long isolation, were tragically vulnerable.[3] About the only ecological tampering left untried was an introduction of Siberian tigers among the bison herds of the Great Plains, but there is little doubt that the consequences, had it been tried, would have been of enormous interest to all parties concerned.

One of the less celebrated of these imported organisms was *Equus asinus*, the donkey (Victorian usage), jack (when employed for breeding), jackass (when audible), ass (biblical and metaphorical), or, in those parts of the United States influenced by the Spanish tradition, burro. He is brown or gray, closely enough related to the horse to interbreed with it but sufficiently distant that the product of the union is almost always sterile. That the burro evolved under desert conditions is shown by his small size, indiscriminate attitude toward anything even remotely edible, keen hearing, intelligence, resistance to heat and sunlight, physical endurance, and ability to scent water over great distances.[4] A close environmental associate is the goat, who takes the high elevations, while the burro prefers the lower flats. Both animals enjoy a moderately bad name in folklore. The goat is stereotyped as ill-smelling and of unbridled lechery, two attributes not undeserved by billy goats of seniority and standing; yet the Spanish word for goat, *cabrón* is a synonym for "cuckold." In Anglo circles the burro is usually equated with social imbecility, a libel probably originating in a fancied resemblance of his loud bray to the hilarity of a rustic of defective reasoning powers and reinforced by its latter-day resemblance to the canned laughter of a televised situation comedy.

First domesticated in the prehistoric Near East, the ass flourished in the Mediterranean world in consequence of his docility, low cost, ability to bear burdens seemingly out of all proportion to his diminutive size, and patent inedibility. Although he can work as a draft animal,[5] he serves best as the poor man's porter. Throughout the ages his brainless cousin the horse has symbolized pride, passion, and war, while the ass has been equated by those who know him best with humility, meekness, and patience in the face of extreme provocation. Literature, beginning with the Old Testament and continuing through the New, is replete with references to the ass, most of them commendatory. The Greco-Roman writer Apuleius composed a work chiefly devoted to the adventures of a human soul trapped by witchcraft in the body of an ass. In medieval times the ass fell from literary grace, possibly because the clergy were required to demonstrate their status by riding asses or mules in lieu of the horses monopolized by the fighting chivalry. The Hohenstaufen emperors were so given to inflicting donkey-centered indignities upon their ungrateful Italian subjects that the works of

Dante are replete with references to a "figging sign," which some commentators think to be a souvenir of this quaint Germanic custom. Worse was to come, however.

The Renaissance authors of popular but unedifying works alluded frequently to the ass or to certain of his physical or moral attributes. Boccaccio, Rabelais, Margaret of Navarre, and the compiler of the *Hundred New Stories* retailed anecdotes in which donkeys, however sordid, usually emerge as models of grace and decorum in comparison with the human protagonists. Since this body of literature was Mediterranean in origin or influence, it seems clear that the ass and its lore were far more institutionalized in that region than in the less congenial climate of northwestern Europe. Chaucer, for instance, has little to say about and no lines spoken by the donkey.

At some unrecorded date experiment established that the foal of an ass and a horse was a large, sturdy, and intelligent animal which combined most of the better features of both species. The hybrid mule was an all-purpose animal which pulled in harness as well as it portered, though its pace and common sense usually disqualified it for direct military deployment. By the age of discovery Spain was held to be the home of the best breeding jacks, the finest mules, and the most vicious and depraved muleteers—a combination traditionally regarded as advantageous and mutually reinforcing. J. Ross Browne observed Hispanic muleteers in action during the rush to Washoe across the Sierra passes in 1859. He wrote:

> The cries and maledictions of the *vaqueros* were perfectly overwhelming; but when the mules stuck fast in the mud, and it became necessary to unpack them, then it was that the *vaqueros* shone out most luminously. They shouted, swore, beat the mules, kicked them, pushed them, swore again; when all these resources failed, tore their hair, and resorted to prayer and meditation. The *vaqueros* made the cliffs resound with their Carambas and Carajas, their Dona Marias and Santa Sofias! a language apparently understood well by mules. The *vaqueros* were in a perfect frenzy of rage and terror combined—shrieking Maladetto! Caramba! and Caraja! till it seemed as if the reverberations must break loose the snow from above, and send an avalanche down on top of us all.[6]

When brought to the Americas, the Spanish donkeys adapted themselves to the land and to the inhabitants thereof. The common burro of the Southwest was far less comely than a Spanish jack of sixteen quarterings of nobility. Centuries of unsupervised inbreeding and elimination of the unfit did to the burro what they did to the Indian mustang, rendering it smaller, shaggier, and less statuesque, while conferring self-reliance and ability to survive mistreatment by man and nature. A missionary visiting Sonora in the eighteenth century noted there that

asses thrive well and are productive in Sonora in spite of the continuous stealing of the Apaches, and for this reason, inasmuch as this animal requires but little for its maintenance, and finds its daily allowance even in the dung-hill of the towns, the natives of the Province value them and at the present time use them largely as beasts of burden.[7]

Following the Moorish tradition, the Mexican rode the burro both as a matter of course and to indicate his status, inasmuch as only an Indio would be seen afoot in public. To this day, in fact, a common sight in a back-country Mexican village is a half-dozen children sticking up like a line of clothespins from the swayed back of a resigned and patient donkey. The animal, incidentally, is mounted over the rump, in a leapfrog manner.[8] Its tallow is regarded as a superior article for the greasing of *carreta* wheels, it is "necked" to wild horses to assist in gentling them, and makes itself generally useful, it is reported, in the manufacture of hot tamales.[9] The burro detected in cornfield pillage, however, was punished by having a portion of one ear lopped off by the sweep of a machete. Such "gotch-eared" animals could thereafter be detected and forestalled, or their owners could be required to give compensation for their thefts.

Observing the habits of a scrub burro, the prospector Daniel E. Conner was moved to comment:

I never knew or heard of one of these sleepy little animals kicking their owner in my life, as badly as I have known some of them to need it. The Mexican addresses it by the name of "burro" as though he was commanding a person and halts the animals by a prolonged "Sh-e-e-e" like one frightening away fowls. The *burro* is a mischievous animal about a camp. The carniverous little brute will steal bacon or other meat to eat like a dog and will almost starve lounging around the camp, in quest of baconrinds and other *debris* in the way of bones &c.

If he ventures away from the camp and becomes frightened he will hasten back for protection and usually when he gets there, he will wallow in the dying embers of the campfire like a lazy negro, until he is scorched all over. When he gets half of the hair burned off him, he will stand over the fire with his nose to the smoke, and his eyes shut, perfectly still, sleepy and happy, for hours at a time or until he is awakened by some one.[10]

If the burro had a natural enemy, it was the puma, or mountain lion. It was noted that in lion country burros grazed during the day a hundred yards or more from camp but that as soon as night fell they tended to crowd up to the fire for human companionship and protection.[11] It has been claimed that the burro's defense was to lie flat on its back, squalling and kicking like a cat itself[12] but that this measure could be adopted only in daylight. At night it would have less chance, and Conner noted an incident in which a badly lion-bitten burro came right into his cabin to elude an attacker.[13]

Whether noble jack or scrub burro, *Equus asinus* is not devoid of tender sentiments, which utilitarian man has put to good use. One of the diplomatic presents received by George Washington shortly before he became president of the United States was a fine jack, Royal Gift, donated by the King of Spain for the improvement of the mules of Mount Vernon. The jack proved sluggish and indifferent toward his duties until July, 1786, at which time Royal Gift suddenly became highly motivated. A year later Washington was boasting, "He never fails," and was charging his neighbors five guineas a session.[14] Royal Gift labored to such good effect that mules soon became very popular for pack transport, mine haulage, drayage, and agricultural work in preindustrial America, and it is said that the legitimate heirs of his body, gotten upon two Maltese jennies of the Washington stud, became the ancestors of the highbred Mammoth jacks, which some have suggested should be declared the state bird of Missouri.

Since mules usually cannot reproduce their kind, the supply of Mammoth jacks for siring purposes had to be numerous and well maintained. As late as the 1930's most agricultural counties included among their residents at least one of these jacks, and notices to that effect were regular institutions in rural weekly newspapers. Headed by a smudgy cut of a quadruped with long ears and a frayed-rope tail, the advertisement informed all interested parties that a jack of pedigree and coat armor was standing at a given barnyard. This stylized language was a convention designed to avoid giving offense to anyone who might object to the more accurate statement that the jack was actively prepared to let his light so shine before men that all could see his good works.

Legal as well as economic history was created by this institution. In a famed Nebraska lawsuit, commonly known as the Celebrated Jackass Case, attorneys wrestled at some length with the issue of express and implied warranty as these concepts related to one jack of defective parts and his aggrieved purchaser. In a noteworthy opinion the learned court held forth:

The evidence before the Court is amply sufficient to establish an express warranty on the part of the defendant that the animal in question was a fit and suitable one for breeding purposes, and the Court therefore finds that there was a warranty. But even conceding that there was no warranty, there surely can be no question, under the evidence, that there was an implied warranty as to its fitness as a breeder and foal-getter. The defendant must surely have known that the plaintiff was purchasing the animal for breeding purposes only. He knew at the time of making the sale that no reasonable man would attempt to use a jackass for any other purpose than to outrage nature by propagating mules. He could not have supposed that the plaintiff desired to acquire a jackass for a pet. The animal is wholly unsuitable for that purpose. Its form is neither pleasing to the eye, nor its voice soothing to the ear. He is neither ornamental in his appearance nor amusing in his habits; he is valuable only as he is able and willing to propagate the mule species.

It appears from the evidence that after purchasing the animal, the plaintiff on several occasions caused him to be placed in the society of certain soft-eyed, sleek-coated young mares, that were in the pink of that condition which is supposed to arouse the interest and attract the attention of any reasonably amorous jackass, but that he passed them up and knew them not. The defendant admits representing to the plaintiff that the jackass in question would do that work for himself. But evidence shows that if he was ever possessed of that valuable and charming accomplishment, he failed on the occasions just mentioned to practise it with the zeal and ardor becoming of an ambitious jackass in full possession of his faculties.

He was, indeed, a worthless, unpedigreed and impotent jackass, without pride of ancestry or hope of posterity—a source of disappointment to his female friends, and an item of expense to his owner. There is no brute in all the animal kingdom more worthless than a Missouri-bred jackass afflicted with lost manhood. He was not as represented and warranted by the defendant, and the plaintiff is entitled to recover.[15]

"The Law," said Dr. Johnson, "is an ass." The learned jurist quoted above, like many of his brothers, may have known something about the law but little of the asinine qualities to which Johnson referred. Reasonable men have purchased or otherwise acquired jackasses for a multitude of other purposes. Not so many years ago burros of Mexican descent were offered for sale as pets by a large mail-order house which, by repute, found many buyers. Farms breeding burros for sale as pets can be found in many parts of the country. Burros lend themselves to this purpose as readily as do Shetland ponies and are not as prone to nip or kick their young proprietors. His honor's ignorance of a jackass's psychology is also demonstrated by a fact well known in agricultural circles: a jack prefers the company of its own kind, and, once exposed to the blandishments of a jenny, it will tend to remain constant to maidens of its own tribe and will refuse to go seeking like Samson after females of foreign parts and false idols. Nevertheless, beneath his veneer of faithfulness and apparent consideration the jack has a dark and brutish streak of male chauvinism, insisting on having his way with a reluctant jennet, deaf to her pleas and entreaties.[16]

The honorable court displayed prima facie ignorance of the fact that the burro was of prime significance in the winning of the West, the humble ass being as much a part of the prospector's equipment as was his pan. The boomers who rushed from California to the Comstock excitement in 1859 almost without exception packed their possessions on burros over the Sierra passes. Once the need for their services had passed, the animals were turned loose. That was grave error, for around a camp burros quickly became as much of a nuisance and for the same reasons as a kleptomaniac Wagnerian soprano. They lurked about the hillsides and flats until the prospectors left camp in the morning. At once the animals descended upon the tents to steal flour, sugar, bacon, beans, woolen shirts, and even the gunnysacks in which salt meat had been stored. When the prospectors returned, the burros would be innocently grazing the hillsides again, oblivious to the hard swearing which greeted the discovery of their depredations. One animal, however, unwisely ate a quantity of self-rising flour, drank copiously of water, and was in the evening found dead, his barrel as round as an apple and his legs standing out like those of a carpenter's bench.[17]

The burro plague moved one "queer genius" of Virginia City to compose a scriptural parody which received wide circulation. There is even the faint possibility that it may have given Sam Clemens the germ of an idea which later bore fruit in *A Connecticut Yankee*, and *1601*. It is quoted by William Wright:

Some Account of Ye Washoe Canary

Let it be proclaimed at the outset that ye Washoe canary is not at all a bird; and though hee hath voice in great volume, lyke unto that of a *prima donna*, yet is hee no sweet singer in Israel. Hee is none other than ye ungainly beaste known in other landes as ye jackass. You may many times observe ye Washoe canary strolling at hys leasure high up on the side of ye craggy hill and in ye declivous place, basking in ye picturesque and charging hys soul wyth ye majestic. Hee rolleth abroad hys poetic eye upon ye beauties of nature; yea, expandeth hys nostryls and drinketh in sublimity.

Hee looketh about hym upon ye rocks and ye sage-bushes; hee beholdeth ye lizard basking in ye sun, and observeth ye gambols of ye horned toad. Straightway hys poetic imagination becometh heated, hee feeleth ye spirit upon him; hee becometh puffed up with ye ardent intensity of hys elevated sensations; hee braceth outwardly hys feet and poureth forth in long-drawn, triumphant gushes hys thunderous notes of rapture, the meanwhile wielding hys tayle up and down in the most wanton manner. Hys musick does not approach unto ye ravishing strains whyche descended through ye charmed mountain of Alfouran, and overflowed with melody the cell of the hermit Sunballad. It hath, in some parts, a quaver more of Chinese harmoniousness.

A wild, uneducated species of canary was thought worthy of mention in ye booke of Job, among the more note-worthy beasts and birds of ye earth; now, how much more worthy of description must be the cultivated and highly accomplished warbler whyche is ye subject of this briefe hystory? We shall presently see that hee will compare favorably with any fowl or beaste of whyche we have mention in ye goode booke. Of ye leviathan we read—"Who can come to him with a double bridle?" But, ah! who dare come to ye Washoe canary wythe a Spanish-bitted double bridle, two rope halters and a lasso? Again, of ye leviathan: "Lay thine hand upon hym, remember the battle, do no more." Verily, I say of ye Washoe canary—lay thine hand upon hym, remember hys heeles, do no more.

Of ye behemoth it is said: "He moveth hys tayle lyke a cedar," but when ye Washoe canary giveth vent to hys sudden inspiration in an impromptu vocal effort he moveth hys tayle like unto two cedars and one pump-handle.

Again, of ye behemoth—"He eateth grass as an ox." Ye Washoe canary not only eateth grass, but in ye wild luxuriance of hys voluptuous fancy, and hys unbounded confidence in hys digestive capacity, rioteth in ye most reckless manner on sage-brush, prickly-pears, thorns and greasewood.

Of ye horse: "He smelleth ye battle afar off and saith, 'ha, ha!'" Now, not any horse can further smell out a thing presumed to be hidden—sugar, bacon, and ye lyke—than ye Washoe canary—then, indeed, hys "yee-haw" far surpasseth the "ha, ha!" of a horse-laugh. What are ye wings of ye peacock or ye feathers of ye ostriche to ye fierceness of hys foretop and ye widespread awfulness of hys ears?

Of ye horse: "He swalloweth ye ground in fierceness and rage." Now, ye Washoe canary swalloweth woolen shirts, old breeches, gunny sacks and dilapidated hoop-skirts when in a state of pensive good nature—what, then, must we suppose hym capable of swallowing, once hys wrath is enkindled and all ye fearful ferocity of hys nature is aroused; Such is ye Washoe canary. Be in haste at no time to proclaim a victory over him.[18]

Despite its poor ability to accommodate to an urban environment, the burro was an ideal working partner for the western prospector, a man who usually had some problems of his own in adjusting to city life. A prospecting outfit and several hundred pounds of food and water while outbound and perhaps nearly as much in ore samples or even surficial high-grade on the return trip required portage. Burros were well suited to such tasks and often enough wild scrubs could be had merely for the taking. The burro lived entirely off the country, however barren, caring for himself as a matter of course. He was at home on the most narrow and precipitous grades—much country explored in the nineteenth century has never been prospected again with improved methods, simply because a burro could easily go where a four-wheel-drive vehicle would stall or turn over. Although his unpredictable bray might attract the attention of unfriendly aborigines, no Indian would consider stealing him.[19] He might kick to pieces a load of which he disapproved, but he would seldom run off with it, leaving the prospector stranded. He required no grain, no picketing, no shoeing, and no coddling beyond the most elementary decencies of livestock management. When the prospector began actively to prospect, he could turn his burros loose and dismiss them from his mind the better to concentrate on the business at hand; when the canaries were again wanted, they could be rounded up with relative ease.

Opinions varied about whether the prospector should ride and, if so,

167

A desert canary and associate. The picture was probably posed and faked to a degree. The pack load is lumpy and unprofessional, while the gun toter is in some danger of a swift kick, well deserved. The burro is, however, probably genuine.

what. Some American desert rats rode large jacks fitted with McClellan saddles. Others preferred horses (a mare or gelding, for jacks and stallions would fight, often to the death[20]). The incorrigible old-timer usually footed it, for it is far easier when afoot to bend over to examine likely float than it is to dismount for that purpose—and those bitten by the gold bug can no more ignore an intriguing rock than a passing feist can ignore a fire hydrant.

Once properly packed and convinced that holidays were over, well-trained burros followed their natural leader in whatever formation best suited the terrain, needing no communal guide rope as did pack horses. This trail leader was usually a jack, but the night leader, whose emblem of office was a neck bell, was invariably a wise old jenny. It was she who assumed responsibility for selecting the feeding ground where the burros grazed most of the night and the bedding ground where they laid up to rest toward morning. In theory the prospector needed only to follow the bell when he wished to reclaim his string and then lead the belled jenny back to camp, whereupon the rest would follow her. In practice she usually chose so secluded a spot for the morning nap and moved about so little that the bell remained silent and the animals invisible. This resulted in the joke that lazy burros would bring the bell jenny feed in their mouths so that she could enjoy a meal and they their rest without ringing the bell that would enable the prospector to summon them to work.

Their good eyesight and intense curiosity made the burros excellent sentries. A man bending over a pan or outcrop was completely absorbed and totally defenseless. Many are the accounts and even daguerreotypes of prospectors' bodies literally filled with arrows, huddled beside their rockers. But a vigorous clanging of the jenny's bell was a good alarm signal, reinforced by the alert stance of her long ears. By following her gaze, the prospector could see who was coming up the trail in time to take such measures as seemed appropriate. When the burros slept, along toward dawn, they gave no such protection, but Indians never attacked at night,[21] and all but greenhorns had trained themselves to roll out of the blankets before first light and assume a defensive stance well away from the camp until the hour of greatest danger had passed.

The lonely prospector was inclined to carry on conversations with his burros, swearing that the wigwag of their ears supplied the other end of the dialogue. On occasion the burro provided other forms of amusement. Herman W. Albert, a young pack prospector in Nevada during Goldfield times, was enormously fond of the high-bred animals he celebrated in sentimental prose, but even he ran out of patience with Jennie B, a jennet of aggravating habits. Having in store a pan of unsuccessful gingerbread dough, he plastered the mixture over the jenny's back and then stood away to enjoy the spectacle:

She stood there the picture of unconcern, as if she meant to ignore the incident. She wasn't fooling me, though. I knew she was literally boiling inside. But whatever plans she was fostering to get even with me were upset by Dan [one of Albert's jacks]. He was right at my heels, and before I realized what he was up to he brushed by me. His big brown ears went bobbing up and down as he greedily lapped up the enticing confection which was beginning to spread down Jennie B's flanks. . . . she let fly at him with everything she had in her

169

lightning heels. Dan, however, merely crowded in closer, smothering her kicks, while his tongue never missed a single lap. Old Gal's [Albert's other jenny] feminine curiosity got the better of her. She must get her finger, rather her tongue, in the pie too. She took up a position on the opposite side, and now Jennie B was caught between two tongues and to imagine she was liking it would have smacked of complete asininity, judging by her humped-up back and flattened ears. Pivot and swing as she might, she couldn't kick herself out of this indignity, much less so when Rickey [Albert's other jack] horned in to see what was cooking and discovered that he had been missing a mighty good thing. He tore in for fair, to make up for lost time.

With three unbridled tongues massaging her spine and ribs Jennie B clearly was on the verge of hysteria. She let out an abortive bray and started to run. . . .

Roughly they were describing a great circle. Dan and Rickey, getting in their licks on the same side, kept forcing the frantic Jennie B inwards toward the vortex against the pressure coming from the other side by the outnumbered Old Gal. Round and round they tore over the rough sage-grown area, one kicking, three licking, and I praying.

Then that which I prayed for happened. Jennie B went down in a heap. She would have been up again in a jiffy, for she was as nimble as a cat, but each time she got her front feet out from under her she was pushed back off balance by the unrestrained impetuosity of her erstwhile pals to whom she now was nothing more than an all-day sucker.[22]

The management of the single burro or of a string of pack animals was both an art and a science. The Mexican provided the burro with a straw-stuffed pad over which the sacks of cargo or faggots of firewood were hitched, it being not uncommon "to see a Mexican driving a donkey with a square load of wood on him that looked in the distance like there might be a half cord of it."[23] Professional American packers preferred a packsaddle cinched around the animal's underside. Loads varied upward from one hundred pounds depending on the size of the animal and the ruggedness of the country, with the packer "throwing" a diamond hitch over the load to hold it securely. Like its relatives the horse and the mule, the burro disliked a tightly cinched packsaddle and while it was being adjusted would swell up like an adder so as to secure as much slack in the cinch as possible. If this was permitted, it would not be long before the load rolled under the belly, where it would promptly be kicked to pieces. The time-honored remedy for such a trick was to wait until the jack was about to explode and then drive a knee sharply up into his belly. As the animal expelled his breath, the cunning packer then tightened the cinch to his own satisfaction. As a concession to comfort, however, each burro had his own tailor-made packsaddle, on the fore end of which his name was carved.

Despite the libels upon his race, the burro, like Uriah the Hittite, was continually in the forefront of the battle. He packed the prospector's equipment and supplies into unknown country and brought out the first ore samples that set off mining excitements. In new camps established in difficult terrain, pack trains of jacks brought in supplies and carried out concentrates.[24] As soon as the trails were sufficiently improved, mules and horses took up the work of packing, only to be supplanted in turn by ox-pulled freight wagons and ultimately narrow-gauge railroads. By the time the shrill whistles of the narrow-gauge engines were echoing in the mountain valleys, the jack was gone—marching beside yet other prospectors heading out where the float was rumored to be rich and the prospects interesting. There are few monuments to the patient, long-eared partner of the mineral finders, but his place in folklore is secure—even if for the wrong reason.

Some obscure psychological quirk has usually impelled the general public (and those who pander to its taste in cheaper literature) to the belief that mineral strikes were seldom the product of anything but pure luck and that this luck was somehow connected with an animal. As early as the tenth century the first big silver strike in the Harz Mountains of Saxony was credited to a horse who, bored and restless, pawed open the ground where his owner had tied him, revealing a ledge of silver-lead ore.[25] Even Georg Bauer, the father of scientific-mining literature, swallowed this story without noticeable difficulty, although Bauer was otherwise a man hard to deceive. Perhaps it is felt that such a strike is the product of mere good fortune somewhat on the order of winning a lottery and that therefore another might as easily fall to any man.

On the American mining frontier this superstition gave rise to the great burro-dowsing myth, much cultivated by writers who know less about prospecting than about editorial prejudices. This myth in substance holds that the real locator of a given mineral strike was the prospector's faithful burro—that the gentle jackass noticed the outcrop, sank on the ore, bagged the samples, put in the location monuments, recorded the location, and occasionally ran the assays before turning the claim over to his lord and master, who awoke from his nap long enough to reward him with the traditional bushel of oats. In testimony whereof are the words of Jim Wardner, a Coeur d'Alene mine promoter, who dwelt upon the feats of Noah S. Kellogg's jackass in discovering the great Bunker Hill and Sullivan silver-lead-zinc lodes in 1885:

> Looking across the creek we saw the jack standing upon the side of the hill, and apparently gazing intently across the canyon at some object which attracted his attention. We went up the slope after him, expecting that, as usual, he would give us a hard chase; but he never moved as we approached. His ears were set forward, his eyes were fixed upon some object, and he seemed wholly absorbed. Reaching his side, we were astounded to find the jackass standing upon a great outcropping of mineralized vein-matter and looking in apparent amazament at the marvelous ore-shoot across the canyon, which then, as you see it now, was reflecting the sun's rays like a mirror.[26]

Thomas A. Rickard, the great mining engineer and editor, snorted with indignation at this palpable fraud. He pointed out at some length that silver-lead exposed to the atmosphere turns rapidly a rusty red or a dull black and concluded acidly, "The talk of a glittering mass of silvery ore sticking out of the mountain-side so brilliantly as to mesmerize the ass, and others not any wiser, is pure moonshine."[27]

As a matter of fact some animals *have* been more or less employed in direct prospecting, whether they willed it or not. The pig (*Sus scrofa*) is trained by the wily Gaul to prospect for truffles, whose peculiar odor Jacques or Hyacinthe can detect beneath several feet of earth. Gopher hills have been experimentally panned in an effort (not always unsuccessful) to see whether a pay streak of placer gold may lie beneath the surface. Not long ago both the Finns and the Russians claimed that dogs were capable of detecting the faint smell of subterranean sulphides decomposing into hydrogen sulphide, their masters taking it from there on the theory that where sulphides exist metallic ores may exist also. Ant hills have been known on rare occasions to display gold colors, particularly after the rains.

Virtually the only American instance of direct prospecting by an animal is reputed to have occurred no earlier than the great uranium rush of the 1950's, in the area of Grants, New Mexico, by one Johnny Gossett and his dog, Butch. This animal, a typical mining-camp dog, was equipped by his owner with a piece of dental X-ray film on which a paperclip had been superimposed, taped securely to the end of the dog's nose. Gossett, Butch, and a portable film-development laboratory drove in a pickup truck to some likely area, whereupon Butch was released, while Gossett took his ease. Butch rambled far and wide over the prospect, sniffing at every interesting shrub or tree trunk, marking it as having been investigated, and then moving on to the next. By report Butch could quarter a twenty-acre prospect in half a day. Upon Butch's return Gossett removed and developed the film on the spot. Should the paper clip leave a sharp image on the film, it was time for Gossett to warm up his Geiger counter and retrace the dog's steps to find the radioactive "hot spot" responsible for the film's reaction. The editor who printed this account of "the first working prototype of the modern, mobile, field geochemical laboratory," put in his own two cents' worth by adding:

Without selling short Johnny's ingenious system of presorting uranium occurrences by remote control, we have just one observation to make. Judging from the activities of a couple of healthy pups belonging to yours truly and which romp daily around Crabgrass Acres, man's best friend is probably best suited for shaft sinking activities or test trenching. At least the general state of destruction around what jokingly passes for the lawn and garden would indicate so.[28]

Notes

Chapter I

1. Henry Hamilton, *The English Brass and Copper Industries to 1800,* 1–44.
2. D. B. Barton, *A History of Tin Mining and Smelting in Cornwall,* 133–36.
3. John R. Leifchild, *Cornwall: Its Mines and Miners,* 299.
4. A. L. Rowse, *The Cousin Jacks: The Cornish in America,* 186, 222.
5. Thomas A. Rickard, *A History of American Mining,* 24.
6. Rowse, *The Cousin Jacks,* 170.
7. *Ibid.,* 359. See also D. B. Barton, "The Cornish Miner in Fact and Fancy," *Essays in Cornish Mining History,* 13–66.
8. Frank A. Crampton, *Deep Enough: A Working Stiff in the Western Mine Camps,* 86–87.
9. Thomas W. Knox, *Underground, or Life Below the Surface,* 78.
10. Thomas A. Rickard, *Retrospect: An Autobiography,* 78–79.
11. Thomas Edwin Farish, *The Gold Hunters of California,* 213.
12. The year is an arbitrary approximation of the point when the first-generation Cornish were leaving the mines, when machine drilling and high explosives were coming in, and when mine boards were beginning to rely upon scientifically trained consultants.
13. An "Irish," or "left-hand," dividend was an assessment levied on mine stockholders to finance new exploration and development. It was usually an open admission that previous expectations had not been realized.
14. Crampton, *Deep Enough,* 42–46.
15. *Ibid.,* 101–105.
16. The figures are only approximations, which assume good but not remarkably fine rock and conditions. Heavy ground, much water, or a very hard or abrasive rock might reduce the advance considerably.
17. Knox, *Underground,* 70.
18. Present-day bootleg mine operators who are not inspected or insured occasionally indulge in a vicious practice they call "high-balling." To save time, one partner loads one side of the face while the other drills at his side. This almost guarantees that sooner or later a loaded hole will be hit by the steel.
19. A. B. Parsons, *The Porphyry Coppers,* 561.
20. For otherwise undocumented quotations and anecdotes I am indebted to my associate Robert Lenon, P.E., of Patagonia, Arizona. See also Otis E Young, Jr., *Western Mining.*

Chapter II

1. Often enough cultural or political considerations prevent the losing side from adopting the weapon which the winner has used so effectively. In this case it required a lifetime to train a longbow archer, and the French had no time to lose. Additionally, the French rulers had no overwhelming desire to see their villeins armed with a weapon capable of defeating their own equipment; *jacqueries*, or peasant uprisings, were fresh in the French memory and had been overcome only with great difficulty even though the peasants were armed with but the rudest of weapons.

2. Many other factors were at work which would ultimately have brought the Turkish empire low; for that, Suleiman himself was so besotted as to terminate the Osmanli succession which had provided his empire with an unbroken line of brilliant sultans. On the other hand, even the temporary overrunning of Germanic Europe, had Vienna fallen, would have produced mischief beyond calculation or repair.

3. Thomas Arthur Rickard, *Man and Metals: A History of Mining in Relation to the Development of Civilization*, II, 520–30.

4. Young, *Western Mining*, 83–85.

5. Efforts to put down unionization in the mines of the American West (1890–1900) or to perpetuate peonage in the Bolivian tin mines (1930–1940) broke down in the face of the workers' knowledge of the extracurricular employment of powder. The Russians, with their vast experience in labor coercion, treat their hard-rock miners gingerly but do not hesitate to clamp down in placering and in salt, mercury, and some forms of coal mining, where the mineral may be won without routine use of explosives.

6. Gösta E. Sandström, *Tunnels*, 280–82. See also H. C. Chellson, "From Gunpowder to Modern Dynamite," *Engineering and Mining Journal*, Vol. CXXXVII, No. 5 (May, 1936), 231ff.

7. Edgar Taylor, "High Explosives and Safety Fuse," *Mining and Scientific Press*, Vol. XCVIII, No. 21 (May 22, 1909), 726.

8. In Arizona. In other states the statutory bell signal varies. Colorado, for instance, uses three bells, which in Arizona is reserved for man hoisting only.

9. Crampton, *Deep Enough*, 104.

10. Samuel L. Clemens [Mark Twain], *Roughing It*, 474.

11. Sandström, *Tunnels*, 135, 155. Steam had many theoretical advantages but was difficult to manage in practice, whereas air is less efficient but so practically effective that it remains supreme in the field.

12. To this day air machines preserve the counterclockwise movement. In modern rock bolting, which secures loose ground in the back of the stope or drift, deep holes are drilled into the ground, and long bolts with expanding nuts are inserted. When tightened up, the bolt is firmly wedged into place, and it in turn holds a large plate or washer flush with the rock surface to anchor it. It is tightened by substituting a socket for the drill steel in the chuck of the machine, fitting the socket over the bolt head, and turning on the air. Since the machine's rotation is counterclockwise and cannot be reversed, such rock bolts and their expanding nuts have left-hand threads.

13. Eliot Lord, *Comstock Mining and Miners*, 335–37.

14. Sandström, *Tunnels*, 141–44.

15. T. A. Burgess, "Explosion in Compressed Air Main," *Mining and Scientific Press*, Vol. XCVII, No. 7 (August 22, 1908), 253.

16. Knox, *Underground*, 71–73.

17. "The Late Terrible Disaster," *Mining and Scientific Press*, Vol. XII, No. 17 (April 21, 1866), 248.

18. William Wright [Dan deQuille], *The Big Bonanza: An Authentic Account of the Discovery, History, and Working of the World-Renowned Comstock Lode of Nevada*, 148.

19. Crampton, *Deep Enough*, 166–71.

20. I wish to express my appreciation to Don Bower and the American West Publishing Company for permission to recast substantially in this chapter material by me which appeared as the article "Fire in the Hole! Evolution and Revolution on the Western Mining Frontier," *American West*, Vol. VII, No. 4 (July, 1970), 15–19.

Chapter III

1. Clemens, "The Aged Pilot Man," *Roughing It*, 369–75, well illustrates camp newspaper humor.

2. Joseph Wasson, *Bodie and Esmerelda*, 24.

3. Crampton, *Deep Enough*, 124–30.

4. Emil W. Billeb, *Mining Camp Days*, 121, 127.

5. Cf. *ibid.*, illustration facing p. 225.

6. Farish, *The Gold Hunters of California*, 82–84.

7. Marcia Rittenhouse Wynn, *Desert Bonanza: The Story of Early Randsburg, Mojave Desert Mining Camp*, 168.

8. *Ibid.*, 161.

9. Farish, *Gold Hunters of California*, 87. This prank may have originated in the old southern joke of "mixing the babies" at rural camp meetings, where the infants were deposited in wagons during the night services. Matters here were complicated — or simplified — by the fact that servants as well as mistresses attended these revivals, and that may have given Mark Twain the idea for the plot of *Puddin'head Wilson.*

10. Billeb, *Mining Camp Days*, 124.

11. Parsons, *The Porphyry Coppers*, 564–5.

12. Billeb, *Mining Camp Days*, illustration facing p. 225.

13. The maximum length of hand steel ordinarily used was seldom more than four feet, since experience demonstrated that blast holes deeper than three feet shot with black powder caused more problems than they solved.

14. Drift-steel ears were wider so as to create optimum capacity for the explosive charge, whereas contest-steel ears were much narrower, inasmuch as the objective was maximum depth, not hole volume.

15. Arthur Lakes, *Prospecting for Gold and Silver in North America*, 280.

16. Mrs. Hugh (Marjorie) Brown, *Lady in Boomtown: Miners and Manners on the Nevada Frontier*, 75.

17. I am indebted to Ross Thomas, of Dolores, Colorado, a onetime "straightaway" contest driller, for outlining in a personal interview the finer rules and procedures of drilling contests in June, 1971 (tape recording in author's possession).

18. Crampton, *Deep Enough*, 77–78.

19. Brown, *Lady in Boomtown*, 75.

20. Interview with Verne McCutchan, Arizona state mine inspector.

21. Atha Albert Richie, "The Real Facts about Those Famous Old Hand-Drilling Contests," *Engineering and Mining Journal*, Vol. CLII, No. 11 (November, 1951), 84ff.

22. I thank Martha Hickey, proprietress and manager of the W H Hotel, Mayer, Arizona, for allowing me in April, 1971, to tour her well-preserved establishment and for her valuable comments on its architecture and domestic economy.

23. Georgius Agricola [Georg Bauer], *De re metallica*, 214.

24. Most of these anecdotes were furnished by Daniel T. Mathisrud, of Lead, South Dakota, whose veracity and sobriety make him an unimpeachable source.

25. A frayed length of stick or rope, used to clean or mop out cuttings from a blast hole.

26. A humorous injunction to steal only high-grade ore.

27. Richard W. Lingenfelter, *The Hardrock Miners*, 21–22.

Chapter IV

1. Lakes, *Prospecting*, 64.

2. "This should be a lesson to adventurers; for a few working miners sometimes obtain in the *first* working of a mine, with little capital, far more than a wealthy company with large appliances can afterward secure." From Leifchild, *Cornwall*, 39.

3. Charles H. Dunning and Edward H. Peplow, Jr., *Rock to Riches: The Story of American Mining . . . Past, Present and Future . . . as Reflected in the Colorful History of Mining in Arizona, the Nation's Greatest Bonanza*, 118; see also *Mining and Scientific Press*, Vol. XXXVII, No. 22 (June 1, 1878), 338, which estimates monthly production at fifty thousand dollars.

4. Patrick Hamilton (comp.), *The Resources of Arizona: A Description of Its Mineral, Farming, Grazing, and Timber Lands; Its Rivers, Mountains, Valleys, and Plains; Its Cities, Towns and Mining Camps; Its Climate and Production; with Brief Sketches of Its Early History, Prehistoric Ruins, Indian Tribes, Spanish Missionaries, Past and Present, etc. etc.*, 110–13; see also *Mining and Scientific Press*, Vol. XXXVIII, No. 26 (June 28, 1879), 413.

5. Dunning and Peplow, *Rock to Riches*, 116–18.

6. Thomas A. Rickard, *The Romance of Mining*, 21–22.

7. *Ibid.*, 28.

8. Cf. Otis E Young, Jr., "The Craft of the Prospector," *Montana: The Magazine of Western History*, Vol. XX, No. 1 (Winter, 1970), 28–39.

9. Quotation from the *Laramie Boomerang*, cited in Clark C. Spence, *Mining Engineers and the American West: The Lace-Boot Brigade, 1849–1933*, 70.

10. Rickard, *The Romance of Mining*, 222.

11. Rickard, *A History of American Mining*, 130.

12. Lakes, *Prospecting*, 167–69.

13. Hamilton, *The Resources of Arizona*, 85. This lode may have been the one located by Brunckow in the early 1850's as the San José, described as one mile northwest of the Cerro Colorado, or Heintzelman, group of claims, east of Tubac on the eastern slopes of the Santa Rita Mountains. Brunckow described the outcrop as three feet thick and going three hundred ounces of silver to the ton; cf. Major S. P. Heintzelman, W. Wrightson, and Edgar Conkling, "Report of Frederick

Brunckow," *Report of the Sonora Exploring and Mining Company*, 16. Brunckow was killed with a drill steel and his body flung down the shaft by a Mexican employee, who fled with the last bullion run of the mine.

14. This was presumably snuff or chewing tobacco, since Schieffelin disclaimed smoking.

15. Hamilton, *The Resources of Arizona*, 121–22.

16. Lakes, *Prospecting*, 242, stated that this particular discovery was the Grand Central, but, in view of Schieffelin's narrative, it was more probably the Lucky Cuss.

17. Lakes, *Prospecting*, 242.

18. *Mining and Scientific Press*, Vol. XXXVIII (June 28, 1879), 413.

19. Edward S. Schieffelin, "History of the Discovery of Tombstone, Arizona, as Told by the Discoverer, Edward Schieffelin," "Arizona: Her Resources and Future Prospects," and "Edward Schieffelin" (MSS P-D2, Bancroft Library, San Francisco). Other information from research paper by Thomas Rykowsky, "History of Mining in Tombstone, Arizona: 1876–1902," 1–19; and B. S. Butler, E. D. Wilson, and C. A. Rasor, *Geology and Ore Deposits of the Tombstone District, Arizona*, 38–50.

20. Lakes, *Prospecting*, 242.

21. Hamilton, *The Resources of Arizona*, 79–85.

Chapter V

1. Diodorus Siculus, *Diodorus of Sicily*, III, 12–14.

2. Sandström, *Tunnels*, 36–37, notes evidence of chicken ladders in the Laurium mines, so installed that they afforded adequate clearance in the center of the shaft for a hoist bucket.

3. This was probably the Mowry Mine, near the present town of Patagonia, Arizona.

4. Knox, *Underground*, 696.

5. Richard S. Kirby, Sidney Withington, Arthur B. Darling, and Frederick G. Kilgour, *Engineering in History*, 97. These authors, like most authorities, rely on the sketch in the Utrecht Psalter of ca. A.D. 850 as the first representation of a crank forged from a single bar of iron.

6. L. Sprague De Camp, *The Ancient Engineers*, 156–57, notes that the ancient grain-grinding quern was crank-actuated and deduced from a bucket-chain bilge pump recovered from the Neronian-period ship of Lake Nemi that it was almost certainly driven by a combination crank and flywheel.

7. Rickard, *Man and Metals*, I, 427–28.

8. Olga W. Smith, *Gold on the Desert*, 126.

9. Wooden-stave sinking buckets persisted for a long time; I observed the ruins of one outside the office of the local newspaper in Rico, Colorado, although I have never seen another anywhere else.

10. Bauer, *De re metallica*, 105, 123.

11. Thomas A. Rickard, *Journeys of Observation*, 192.

12. Bauer, *De re metallica*, 154.

13. *Ibid.*, 199.

14. Young, *Western Mining*, 55ff.

15. Bauer, *De re metallica*, 322–23.

16. Rickard, *Journeys of Observation*, 188–95.

17. Actually the Cornish customarily worked not in timed shifts but in "cores," or spells, which were based on a quota of ore removed or feet driven. On the average, however, the time consumed in meeting these quotas approximated a shift.

18. Louis Simonin, *La vie souterrane*, 237, 239, 243ff.

19. In this period marine anchor hawsers of comparable size and strength were handled by several ingenious methods too complicated to describe, which tended to eliminate altogether the breakage of strands which would result from running them over circular guides or drums. Mooring hawsers were in use only a small fraction of the time, however, in contrast to mining, where employment would be constant and the sailors' roundabout handling methods unfeasible.

20. John A. Roebling, "Manufacture of Wire Ropes," in Samuel Rapport and Helen Wright (eds.), *Engineering*, 157–62.

21. Lord, *Comstock Mining and Miners*, 224.

22. For a photograph of a counterbalanced wire belt hoist, see Billeb, *Mining Camp Days*, 99.

23. John Hays Hammond, *The Autobiography of John Hays Hammond*, II, 493.

24. Lord, *Comstock Mining and Miners*, 402–403.

25. James V. Thompson, "The Visitation Engineer writes again," *Metal Mining & Processing* (September, 1964), 41–43.

26. *Ibid.*

27. Because of the average sixty-degree dip of the formations in the Mother Lode gold mines of California in the Grass Valley-Nevada City region and because of their great depth (up to ten thousand feet on the dip), skips installed in inclines were almost exclusively employed for all hoisting in that district. Men rode in special "man skips," but conventional manways and cages were installed wherever possible because of safety considerations. Cf. Jack R. Wagner, *Gold Mines of California*.

28. The hoist rope was attached to the high point of the bucket bail also, but the bail was capable of rotating in the sockets (somewhat below the bucket rim) into which the ends fitted. Miners sitting on the rim of an empty bucket were thus above its center of gravity and could be dumped out by a moment's carelessness.

29. Wright, *The Big Bonanza*, 225.

30. *Ibid.*, 120–22.

31. Rickard, "Across the San Juan Mountains," *Journeys of Observation*, 60–66. See also Hammond, *Autobiography*, II, 520–22.

Chapter VI

1. J. Ross Browne, *Report of the Mineral Resources of the States and Territories West of the Rocky Mountains*, 454.

2. Thomas Edwin Farish, *History of Arizona*, II, 211.

3. Hiram C. Hodge, *Arizona as It Is; or the Coming Country, Compiled from Notes of Travel During the Years 1874, 1875, and 1876*, 92.

4. Farish, *History of Arizona*, I, 298.

5. Rossiter W. Raymond, *Statistics of Mines and Mining in the States and Territories West of the Rocky Mountains*, 257.

6. [Writers' Program of Arizona, WPA], *Arizona: A State Guide*, 57, offers the burro-discovery thesis, at which none of the earlier references even hint. In point of fact, the Walker Party had stumbled across the Vulture outcrop in the spring of 1863 while following a gulch which they mistook for the main course of

the Hassayampa, but none of its members recognized its possibilities. Cf. Daniel Ellis Conner, *Joseph Reddeford Walker and the Arizona Adventure* (ed. by Donald J. Berthrong and Odessa Davenport), 85.

7. Farish, *History of Arizona*, II, 212–13, suggests that Wickenburg first located the Vulture on his journey north in the spring of 1863 and was later sued by Van Bibber's assignees. This seems improbable, since, once Wickenburg had found his outcrop, he never left the vicinity again, but he was sued unsuccessfully by one Murray.

8. Raymond, *Statistics*, 257–60; Browne, *Report*, 477–78.

9. *Ibid.*

10. Hodge, *Arizona as It Is*, 92. Some reports suggest that two or three other men assisted Wickenburg and Moore in this original development. Since no mining district had yet been organized in the locality, however, Wickenburg may have had to locate the Vulture as a series of placer claims, paying "dummies" to register a total of five claims, each of 100 feet. This would account for the peculiar division of the Vulture into fifths.

11. Ward R. Adams, *History of Arizona* (ed. by Richard E. Sloan), I, 512.

12. Farish, *History of Arizona*, II, 213.

13. *Ibid.*, II, 212–13. Browne, *Report*, 477–78, states that the Smith and Brill partnership, working the remaining fifth of the Vulture footage between November 1, 1866, and September 1, 1867, mined and milled 4,834 tons averaging thirty dollars a ton.

14. Farish, *History of Arizona*, I, 298, calls this promoter Phillips instead of Phelps, but in any event he had no connection with the Phelps Dodge Company. Genung was ordinarily a rancher in the Peeples Valley and was a good man with a gun as well as with a gold development, for the (Prescott) *Arizona Miner* of November 15, 1879, reported that Genung had elevated the moral tone of the territory by killing Oscar Bear, a local bully. Genung was freed immediately following a brief hearing at Prescott, the territorial capital.

15. The territorial governor, R. C. McCormick, was less restrained. In his legislative message of 1868 he referred to the Vulture as "the Comstock of Arizona." Cf. Farish, *History of Arizona*, V, 37. Needless to say, the mining men were right, and the governor was wrong, unless he was indulging in sarcasm.

16. Raymond, *Statistics*, 257. Raymond and Browne agree that gross production at the Vulture in the six years from 1867 to 1872 was $2.5 million from 118,000 tons of $21 ore. Production costs probably approached $2 million, however.

17. *Ibid.*, 258–59; Eldred D. Wilson, J. B. Cunningham, and G. M. Butler, *Arizona Gold Lodes and Gold Mining*, 161–62; H. H. Bancroft, *History of Arizona and New Mexico*, 587, note 6. In 1868 the Vulture Company grossed $254,000 on $24 ore, but its cost breakdown is illuminating: $4.12 for mining, $8 for freighting to the mill, and $2.81 for milling. Milling losses are not given, but probably approached $5.

18. Farish, *History of Arizona*, II, 214. Sexton's peculations cast a different light on the story given by Rickard, that the Vulture's problems were caused by the miners high-grading ore. While high-grading in thirty-dollar ore is not unknown, it is sufficiently unusual to make the Sexton story the more probable. Rickard, *A History of American Mining*, 270.

19. The Vulture was to go through many more sets of hands, including those of H. A. W. Tabor, and many more stages of development before it was finally abandoned about 1940. Those interested in its history are referred to Duane A. Smith,

"The Vulture Mine: Arizona's Golden Mirage," *Arizona and the West*, Vol. XIV, No. 3 (Autumn, 1972), 231–52. This is a very well researched account, but I take issue with various conclusions regarding reliability of certain sources and other matters of relatively small account. So much error and inaccuracy has been wantonly hung about the Vulture Mine that the article is welcome as a serious and scholarly attempt to get at the truth.

20. William H. Emmons, *Gold Deposits of the World, with a Section on Prospecting*, 170; Waldemar Lindgren, *Mineral Deposits*, 491–95.

21. Gold also combines chemically with the chloride and cyanide ions to form compounds so soluble that they are never naturally found in place. Lode gold may be alloyed with silver percentages and in placers (rarely) amalgamated with mercury. In pyritic and selenitic ores the gold is "carried" or physically commingled with the other metal salts, but microscopic examination of these ores will show that the particles of gold may enclose or be enclosed by the various pyrites or selenides but that the particles themselves (unless telluric) are native.

22. William S. Greever, *The Bonanza West: The Story of the Western Mining Rushes, 1848–1900*, 206. When intruded and slowly cooled at moderate depth, phonolite hardens into a microcrystalline phase whose shards ring like a bell when struck—hence the name. Having low proportions of silica, its presence is unfavorable to mineralization, its color mud brown, and its general utility nil.

23. Marshall Sprague, *Money Mountain: The Story of Cripple Creek Gold*, 11–12. Despite its employment as a source by some historians, this book is extremely weak in many respects. Accordingly, it is cited only with strong reservations.

24. Frank Waters, *Midas of the Rockies*, 117–18. Waters' account is slightly stronger than Sprague's but displays many of the same fundamental weaknesses.

25. Sprague, *Money Mountain*, 12. "White iron" is a common term for arsenopyrite, worthless in small quantities but in massive form a possible carrier of more valuable mineralization. Why H. T. Wood, if sufficiently educated to be a USGS employee, neglected to have it assayed is an open question. Needless to say, the mineral was not arsenopyrite but calaverite.

26. Greever, *Bonanza West*, 206.

27. Sprague, *Money Mountain*, 36–40. See also Rickard, *A History of American Mining*, 143.

28. Sprague, *Money Mountain*, 40–47. It is stated that De la Vergne obtained evidence of calaverite from a "bronze yellow ore," which might mean massive iron or iron-copper pyrites. However, Rickard (*History of American Mining*, 144–45) suggested that the calaverite was first identified by Richard Pearce, a mineralogist of Denver, or by an unnamed prospector who built a campfire in trachyte carrying calaverite and thus virtually had the identification forced on him.

29. Waters, *Midas of the Rockies*, 119–24, offers accounts of Stratton's strike which are not denied to be apocryphal, the products of barroom oral history. Since they differ considerably from the account in Rickard, *A History of American Mining*, 145–46, reliance has been placed on Rickard.

30. Rickard, *A History of American Mining*, 145–46.

31. Sprague, *Money Mountain*, 99–107.

32. Lindgren, *Mineral Deposits*, 492. The Roosevelt Tunnel drained the breccia plug at the two-thousand-foot level. It and two shorter, higher predecessors holed through the granite footwall to begin the drainage process long before its official completion date, however. Some stories that the Roosevelt Tunnel encountered

great technical difficulties should be discounted; much of its length was driven through solid granite, almost ideal ground for such a work.

33. Sprague, *Money Mountain*, 224–27.

34. Hammond, *Autobiography*, II, 489–90.

35. Russell R. Elliott, *Nevada's Twentieth Century Mining Boom: Tonopah, Goldfield, Ely*, 4–8.

36. Henry Curtis Morris, *Desert Gold and Total Prospecting*, 8.

37. Lindgren, *Mineral Deposits*, 510–11. See also Frederick Leslie Ransome, *Preliminary Account of Goldfield, Bullfrog, and Other Mining Districts in Southern Nevada*, 11ff.

38. Hope springs eternal among optimists who never cease attempting to get rich in placering such fields of gold float in desert country using mechanical dry-washing machines, which are sometimes of great size and complexity. They operate on the principle of winnowing the gold dust by means of an air blast. In theory they should work very well, but in operation it is soon discovered that the placer sands must be absolutely dry for them to succeed. Even in the most arid country, unfortunately, the sands a few inches below the surface always have just enough moisture adhering to each grain to clog the machine. The operator then installs a big heater to dry his feed and does well in his own sight until he compares his cleanup with his power and fuel bills. At this point one more dry washer is abandoned.

39. Morris, *Desert Gold and Total Prospecting*, 6.

40. After a tour of the Florence Mine headworks at Goldfield, the operators, Mr. and Mrs. Martin Duffy, showed me a specimen of Goldfield high-grade ore which had been polished and set in a brooch. It appeared to be brown until it was inspected closely with the aid of a strong light, whereupon the innumerable pinpoint grains of gold shimmered yellow on the background of the matrix.

41. This should cast some light upon the murky subject of "secret" mines. The Combination strike was so high grade, so easily worked at grass roots, and so novel in kind that the prospectors could readily have kept the good news to themselves had they so desired. Obviously they felt, as had Butler at Tonopah, that wide publicity was far more to their advantage, as indeed it proved to be. The general conclusion is that most of the secret mines of popular lore existed only in someone's imagination and that a small minority were but high-grade point deposits which gave up one load of ore before they gave up the ghost.

42. It might appear unreasonable that percolating hydrothermals should prefer to ascend through the middle of a solid stratum instead of following the contact. The contact, however, would tend to be "welded" tightly by the heat of the over-running flow, whereas later cooling and internal contraction would produce central fissures and joints within the new stratum itself. This phenomenon is familiar to foundrymen, who have to take great pains to prevent such differential cooling in large castings. This is true, however, only of igneous strata; sedimentary strata react in the opposite fashion, the contact remaining relatively open but the stratum itself compacting substantially from ground movement or overburden load.

43. Lindgren, *Mineral Deposits*, 510–11.

44. Elliott, *Nevada's Twentieth-Century Mining Boom*, 16–17.

45. Thomas A. Rickard, "Rich Ore and Its Moral Effects," *Mining and Scientific Press*, Vol. XCVI (June 6, 1908), 774–75.

46. Lindgren, *Mineral Deposits*, 512.

47. Elliott, *Nevada's Twentieth-Century Mining Boom*, 159–60.

Chapter VII

1. Alfred W. Bruce, *The Steam Locomotive in America*, 40–42, 66. Bruce gives a total of sixteen thousand locomotives in the United States in 1860, of which I infer about half were of the "American" type.

2. Cf. Carroll E. Pursell, Jr., *Early Stationary Engines in America: A Study in the Migration of a Technology*. The study is limited to mill engines; Cornish beam engines are ignored.

3. Perhaps the best poem of steam is the now almost forgotten "M'Andrew's Hymn," by Rudyard Kipling.

4. H. W. Dickinson, *A Short History of the Steam Engine*, 24–26. The description given in the text is that of the last and most effective Savery pump.

5. To emphasize the fact that the steam did no direct work, it may be suggested that a vacuum pump (had Newcomen possessed one) could have been substituted for the steam and condensing spray without affecting the principle or the operation of his engine. The sole function of the steam was to provide a space-filling vapor which could be rapidly condensed to produce the desired vacuum.

6. Dickinson, *A Short History of the Steam Engine*, 49–65.

7. Leifchild, *Cornwall*, 182–93. Watt added auxiliary systems: the condenser air pump; the "parallel motion," which eased the torsional strain on the cylinder rod caused by the beam's segment-of-arc travel; and an ingenious clepsydra, which halted the engine at the end of each stroke for the two or three seconds necessary for the pump valves to reverse and reseat themselves. These were more cures for operational "bugs" than basic advances in the art, however.

8. Watt himself always deplored high pressure and direct action, correctly fearing boiler explosions. His fears were amply borne out, especially in the United States, where exploding locomotives and Mississippi side-wheelers were epidemic in the 1850's. In the long run, however, condensers proved so delicate and high pressure so advantageous that the condenser was employed only where conservation of boiler feedwater was an important economic factor, as in marine-engine plants or the present-day central-generation turbine stations.

9. Compounding couples a small, high-pressure cylinder with a much larger low-pressure cylinder whereby the high-pressure exhaust, still carrying much energy, is fed through the low-pressure inlet valve. When correctly calculated and some reheating provided in an "economizer," compounding can be carried on for three stages, and is both powerful and economical as reciprocating steam engines go. Its most familiar American application was in the Mallet freight locomotives, whose characteristic double chuffing was familiar to an entire generation.

10. Erecting a heavy masonry bob wall at the very edge of a shaft was a virtual invitation to trouble unless the ground was extremely strong. Occasionally trouble developed, and the entire front wall of the engine house collapsed into the shaft. American practice has always been to keep machinery well back from the shaft to avoid the twin problems of dead weight and mechanical vibration in this inherently sensitive area.

11. Except for specifically cited references, the narrative to this point has been extracted from the excellent study by D. B. Barton, *The Cornish Beam Engine: A Survey of Its History and Development in the Mines of Cornwall and Devon from Before 1800 to the Present Day, with Something of Its Use Elsewhere in Britain and Abroad*.

12. Leifchild, *Cornwall*, 161.

13. *Ibid.*, 255–56.

14. Bryan Earl, *Cornish Mining: The Techniques of Metal Mining in the West of England, Past and Present*, 36–42.

15. Dickinson, *A Short History of the Steam Engine*, 108–109.

16. Grant H. Smith, *The History of the Comstock Lode, 1850–1920*, 16.

17. Barton, *The Cornish Beam Engine*, 61–62.

18. Young, *Western Mining*, 195–98.

19. Doubling the size of an object multiplies its mass by the cube of dimensional increase (two by two by two), while halving its size diminishes the mass by the cube root.

20. Shallow-draft sternwheel steamboats could carry materials up the Colorado River from its delta at the head of the Gulf of California to the river landings at Yuma, La Paz, and Ehrenberg, but the gain over unloading at Guaymas was relatively negligible.

21. In 1869 the narrow-gauge Virginia & Truckee Railroad was completed between Virginia and the Central Pacific main line at Reno, Nevada, but this happy event did not occur until a decade after the Comstock excitement began.

22. Water-powered chain bailing pumps were no monopoly of the Chinese, being used the world over, but apparently they were first employed in California by Chinese placermen, and so the name stuck.

23. Lord, *Comstock Mining and Miners*, 88.

24. Smith, *The History of the Comstock Lode*, 99–100. It was at this time that Adolph H. J. Sutro conceived the idea of a single great drainage adit to be driven beneath the eastern Rose Peak Range from the Carson River valley, a distance of over four miles. Intrigue delayed the completion of this work until July 8, 1878, almost too late to be of economic service to the district.

25. Elizabeth L. Egenhoff, "The Cornish Pump," *Mineral Information Service of the California Division of Mines and Geology*, Vol. XX, Nos. 6–8 (June–August, 1967), 60.

26. Clarence W. King, *United States Geological Exploration of the Fortieth Parallel*, III, 124–31, and Plate XVI facing p. 146.

27. Wright, *The Big Bonanza*, 168–73. It was usually unfeasible to employ the sump water from these mines since it was so hot and corrosive that it would destroy any boiler or condenser in a matter of hours. In 1873 a major aqueduct was completed to Virginia City from Marlett Lake in the Sierra Nevada Mountains, and engines installed after this date could use condensers. However, the C. & C. (Consolidated Virginia and California) pump lifted two thousand tons of water a day which was neither acid nor greatly heated, and in 1878 this water, after being cooled by a twenty-foot fall, was used in the engine condenser. See *Mining and Scientific Press*, Vol. XXXVII, No. 26 (June 29, 1878), 408.

28. King, *USGS Exploration of the Fortieth Parallel*, III, 128, states that the Gould & Curry Mine pump was equipped with reduction gears in the manner of the Savage Mine pump described. See also John A. Church, *The Comstock Lode: Its Formation and History*, 25–27. Church noted that the sump water was drained down Gold Canyon and Six-Mile Canyon to the Carson River, its flow being used to carry down mill tailings.

29. Because of the variation of strains imposed upon it, as well as the need to have the teeth in reasonably precise alignment and spacing on its circumference, bull gears had to be cast en bloc.

30. Egenhoff, "The Cornish Pump," *Mineral Information Service of the California Division of Mines and Geology,* Vol. XX, Nos. 6–8 (June–August, 1967), 57.

31. I infer that Attwood meant the reduction gear castings, not the cylinder. Gear castings could be hand-filed to necessary tolerances, but a steam cylinder had to be machined internally by quite massive and specialized lathes. Since it would be easier and far cheaper to bring in a used mill engine than to import machine tools for one engine job, the conclusion seems obvious. It is not suggested, however, that the drive depicted by Mathis is the original installation.

32. Barton, *The Cornish Beam Engine,* 92–96.

33. *Ibid.,* 20, note.

34. Egenhoff, "The Cornish Pump," *Mineral Information Service of the California Division of Mines and Geology,* Vol. XX, Nos. 6–8 (June–August, 1967), 67.

35. Smith, *History of the Comstock Lode,* 245.

36. *Ibid.,* 278–80. Two photographs of the "Union" engine are in Lord, *Comstock Mining and Miners,* following p. 346. The great balance wheel is evidently built up, following western practice. A slightly smaller installation at the Yellow Jacket Mine is described in some detail in *Mining and Scientific Press,* Vol. XXXVI, No. 9 (March 2, 1878), 136.

Chapter VIII

1. The striped skunk, after surgical attention, may become a house pet and suppressor of vermin. The raccoon, if one can abide its thieving ways, will also lend itself to domestication. In the desert Southwest the ring-tailed cat of raccoon antecedents is known as the "miner's cat" from its habit of ingratiating itself with solitary prospectors. None of these, however, may properly be deemed domesticated in the economic sense of the word.

2. It is perhaps significant that the Amerind referred to the "beaver people" and to the "bear people" and regarded the kill of a grizzly as being fully as prestigious as counting coup on an armed and alert human enemy.

3. As far as can be ascertained, the American Indian had no contagious diseases before the coming of the Europeans, with the exception of syphilis, supposedly picked up by Columbus' sailors in the West Indies. However, the close resemblance of *Treponema palledum* to the organism which produces the African yaws and the fact that the Portuguese were in 1493 or thereabouts exploring this area lends weight to the supposition that syphilis was really a pernicious variant of yaws.

4. Frances F. Victor, *The River of the West,* 145.

5. J. Frank Dobie, *The Mustangs,* 218.

6. J. Ross Browne, *A Peep at Washoe, or Sketch of Adventure in Virginia City,* 47, 136, 138.

7. [Anonymous], "Rudo Ensayo [Sonora]" (trans. by Eusebio Guitéras), *Records of the American Catholic Historical Society of Philadelphia,* Vol. V, No. 2 (June, 1894), 141.

8. Conner, *Joseph Reddeford Walker*, 15, 18.

9. Dobie, *The Mustangs*, 192, 232–33.

10. Conner, *Joseph Reddeford Walker*, 143–44.

11. John R. Cook, *The Border and the Buffalo: An Untold Story of the South-west Plains* (ed. by M. M. Quaife), 70.

12. A. F. Tschiffely, *Tschiffely's Ride*, 40–41.

13. Conner, *Joseph Reddeford Walker*, 144–45.

14. Douglas Southall Freeman, *George Washington: A Biography*, VI, 58–59.

15. 121 Nebr. App. 72.

16. Glen R. "Slim" Ellison, *Cowboys Under the Mogollon Rim*, 232.

17. Wright, *The Big Bonanza*, 72–73.

18. *Ibid.* The temptation is strong to attribute "Ye Washoe Canary" to Clemens as an early work, but he did not arrive at Virginia City until more than two years later. In addition, his scriptural-historical parodies were the product of his later years in Hartford. And, regarded subjectively, "Ye Washoe Canary" seems just not good enough to be Clemens' product, apart from the fact that Wright would probably not refer to his journalistic comrade as a "queer genius."

19. Dobie, *The Mustangs*, 40.

20. Walker D. Wyman, *The Wild Horse of the West*, 261–62.

21. It is reported that the Indians assigned religious scruples for the reluctance to fight at night. However, in a night fight casualties are almost always proportional to numbers engaged. Since the basic principle of Indian tactics was to avoid *any* casualties to one's own side, it seems reasonable that in ages past experience had shown them that night fighting did not pay. The conventional Indian attack was therefore limited to the dawn rush, the ambush, and the stalk.

22. Herman W. Albert, *Odyssey of a Desert Prospector*, 74–75.

23. Conner, *Joseph Reddeford Walker*, 15.

24. George W. Stokes and Howard R. Driggs, *Deadwood Gold: A Story of the Black Hills*, 95–96.

25. Rickard, *Man and Metals*, II, 519–20.

26. Rickard, *History of American Mining*, 321–22.

27. *Ibid.*

28. "Behind the By-Lines," *Engineering and Mining Journal*, Vol. CLXVIII, No. 8 (August, 1967), 4.

Bibliography

Books

Adams, Ward R. *History of Arizona*. Ed. by Richard E. Sloane. 4 vols. Phoenix, 1930.

Agricola, Geogius [Georg Bauer]. *De re metallica*. Trans. and ed. by Herbert Clark Hoover and Lou Henry Hoover. New York, 1950.

Albert, Herman W. *Odyssey of a Desert Prospector*. Norman, 1967.

Bancroft, Hubert Howe. *History of Arizona and New Mexico, 1530–1888*. San Francisco, 1889.

Barton, D. B. *The Cornish Beam Engine; A Survey of Its History and Development in the Mines of Cornwall and Devon from Before 1800 to the Present Day, with Something of Its Use Elsewhere in Britain and Abroad*. New ed. Truro, Cornwall, 1966.

———. *Essays in Cornish Mining History*. Truro, Cornwall, 1968.

———. *A History of Tin Mining and Smelting in Cornwall*. Marazion and Penzance, Cornwall, 1967.

Billeb, Emil W. *Mining Camp Days*. Berkeley, 1968.

Brown, Mrs. Hugh (Marjorie). *Lady in Boomtown: Miners and Manners on the Nevada Frontier*. Berkeley, 1968.

Browne, J. Ross. *A Peep at Washoe, or Sketch of Adventure in Virginia City*. Reissue ed. Palo Alto, 1968.

———. *Report on the Mineral Resources of the States and Territories West of the Rocky Mountains*. Washington, D.C., 1868.

Bruce, Alfred W. *The Steam Locomotive in America*. New York, 1952.

Butler, B. S., E. D. Wilson, and C. A. Rasor. *Geology and Ore Deposits of the Tombstone District, Arizona*. Tucson, 1938.

Church, John A. *The Comstock Lode: Its Formation and History*. New York, 1879.

Clemens, Samuel L. [Mark Twain]. *Roughing It*. Hartford, 1883.

Conner, Daniel Ellis. *Joseph Reddeford Walker and the Arizona Adventure*. Ed. by Donald J. Berthrong and Odessa Davenport. Norman, 1956.

Cook, John R. *The Border and the Buffalo: An Untold Story of the Southwest Plains.* Ed. by M. M. Quaife. Chicago, 1938.

Crampton, Frank A. *Deep Enough: A Working Stiff in the Western Mine Camps.* Denver, 1956.

DeCamp, L. Sprague. *The Ancient Engineers.* Garden City, N.Y., 1963.

Dickinson, H. W. *A Short History of the Steam Engine.* London, 1963.

Diodorus Sicilus. *Diodorus of Sicily.* Trans. by C. H. Oldfather. 12 vols. Cambridge, Mass., and London, 1960.

Dobie, J. Frank. *The Mustangs.* Boston, 1952.

Dunning, Charles H., and Edward H. Peplow, Jr. *Rock to Riches: The Story of American Mining . . . Past, Present and Future . . . as Reflected in the colorful History of Mining in Arizona, the Nation's Greatest Bonanza.* Phoenix, 1959.

Earl, Bryan. *Cornish Mining: The Techniques of Metal Mining in the West of England, Past and Present.* Truro, Cornwall, 1968.

Elliott, Russell R. *Nevada's Twentieth-Century Mining Boom: Tonopah, Goldfield, Ely.* Reno, 1966.

Ellison, Glen R. "Slim." *Cowboys Under the Mogollon Rim.* Tucson, 1968.

Emmons, William H. *Gold Deposits of the World, with a Section on Prospecting.* New York and London, 1937.

Farish, Thomas Edwin. *The Gold Hunters of California.* Chicago, 1904.

––––––. *History of Arizona.* 8 vols. Phoenix, 1915–18.

Freeman, Douglas Southall. *George Washington: A Biography.* 7 vols. New York, 1948–57.

Greever, William S. *The Bonanza West: The Story of the Western Mining Rushes, 1848–1900.* Norman, 1963.

Hamilton, Henry. *The English Brass and Copper Industries to 1800.* 2d ed. New York, 1967.

Hamilton, Patrick, comp. *The Resources of Arizona: A Description of Its Mineral, Farming, Grazing, and Timber Lands; Its Rivers, Mountains, Valleys, and Plains; Its Cities, Towns and Mining Camps: Its Climate and Production; with Brief Sketches of Its Early History, Prehistoric Ruins, Indian Tribes, Spanish Missionaries, Past and Present, etc. etc.* 2d. ed. San Francisco, 1883.

Hammond, John Hays. *The Autobiography of John Hays Hammond.* 2 vols. New York, 1935.

Heintzelman, Major S. P., W. Wrightson, and Edgar Conkling. *Report of the Sonora Exploring and Mining Company.* Cincinnati, 1856.

Hodge, Hiram C. *Arizona as It Is; or the Coming Country, Compiled from Notes of Travel During the Years 1874, 1875, and 1876.* Reissue ed.

Chicago, 1965.

King, Clarence W. *United States Geological Exploration of the Fortieth Parallel*. 5 vols. Washington, D.C., 1870.

Kirby, Richard S., Sidney Withington, Arthur B. Darling, and Frederick G. Kilgour. *Engineering in History*. New York, Toronto, and London, 1956.

Knox, Thomas W. *Underground, or Life Below the Surface*. Hartford, 1874.

Lakes, Arthur. *Prospecting for Gold and Silver in North America*. 2d ed. Scranton, 1896.

Leifchild, John R. *Cornwall: Its Mines and Miners*. Facsimile ed. New York, 1968. (Original ed., London, 1857.)

Lindgren, Waldemar. *Mineral Deposits*. 4th ed. New York and London, 1933.

Lingenfelter, Richard W. *The Hardrock Miners*. Berkeley, and Los Angeles, 1974.

Lord, Eliot. *Comstock Mining and Miners*. Reissue ed. Berkeley, 1959.

Morris, Henry Curtis. *Desert Gold and Total Prospecting*. Washington, D.C., 1955.

Parsons, A. B. *The Porphyry Coppers*. New York, 1933.

Pursell, Carroll E. *Early Stationary Engines in America: A Study in the Migration of a Technology*. Washington, D.C., 1969.

Ransome, Frederick Leslie. *Preliminary Account of Goldfield, Bullfrog, and Other Mining Districts in Southern Nevada*. Washington, D.C., 1907.

Rapport, Samuel, and Helen Wright, eds. *Engineering*. New York, 1964.

Raymond, Rossiter W. *Statistics of Mines and Mining in the States and Territories West of the Rocky Mountains*. Washington, D.C., 1872.

Rickard, Thomas A. *A History of American Mining*. New York, 1932.

———. *Journeys of Observation* [including *Across the San Juan Mountains*]. San Francisco, 1907.

———. *Man and Metals: A History of Mining in Relation to the Development of Civilization*. 2 vols. New York and London, 1932.

———. *Retrospect: An Autobiography*. New York and London, 1937.

———. *The Romance of Mining*. Toronto, 1945.

Rowse, A. L. *The Cousin Jacks: The Cornish in America*. New York, 1969.

Sandström, Gösta E. *Tunnels*. New York, Chicago, and San Francisco, 1963.

Simonin, Louis. *La vie souterrane*. Paris, 1867.

Smith, Grant H. *The History of the Comstock Lode, 1850–1920*. Reno, 1943.

Smith, Olga W. *Gold on the Desert*. Albuquerque, 1956.

Spence, Clark C. *Mining Engineers and the American West: The Lace-Boot Brigade, 1849–1933*. New Haven and London, 1970.

Sprague, Marshall. *Money Mountain: The Story of Cripple Creek Gold*. New

York, 1971.

Stokes, George W., and Howard R. Driggs. *Deadwood Gold: A Story of the Black Hills*. New York and Chicago, 1926.

Storms, William H. *Timbering and Mining: A Treatise on Practical American Methods*. 4th ed. New York and London, 1909.

Tschiffely, A. F. *Tschiffely's Ride*. New York, 1933.

Victor, Frances F. *The River of the West*. Hartford, 1870.

Wagner, Jack R. *Gold Mines of California*. Berkeley, 1970.

Wasson, Joseph. *Bodie and Esmeralda*. Facsimile ed. Saratoga and San Jose, Calif., 1963.

Waters, Frank. *Midas of the Rockies*. Stratton Centennial ed. Denver, 1949.

Wilson, Eldred D., J. B. Cunningham, and G. M. Butler. *Arizona Gold Lodes and Gold Mining*. Tucson, 1967.

Wright, William [Dan deQuille]. *The Big Bonanza: An Authentic Account of the Discovery, History, and Working of the World-Renowned Comstock Lode of Nevada*. Reissue ed. New York, 1964.

[Writers' Program of Arizona, WPA]. *Arizona: A State Guide*. New York, 1940.

Wyman, Walker D. *The Wild Horse of the West*. Caldwell, Idaho, 1945.

Wynn, Marcia Rittenhouse. *Desert Bonanza: The Story of Early Randsburg, Mojave Mining Camp*. Glendale, 1963.

Young, Otis E, Jr. *Western Mining: An Informal Account of Precious-Metals Prospecting, Placering, Lode Mining, and Milling on the American Frontier from Spanish Times to 1893*. Norman, 1970.

Periodicals and Newspapers

[Anonymous]. "Rudo Ensayo [Sonora]," trans. by Eusebio Guitéras, *Records of the American Catholic Historical Society of Philadelphia*, Vol. V, No. 2 (1894).

Arizona Miner (Prescott, Arizona Territory).

Burgess, T. A. "Explosion in Compressed Air Main," *Mining and Scientific Press*, Vol. XCVII (1908), 253.

Chellson, H. C. "From Gunpowder to Modern Dynamite," *Engineering and Mining Journal*, Vol. CXXXVII (1936), 231ff.

Egenhoff, Elizabeth L. "The Cornish Pump," *Mineral Information Service of the California Division of Mines and Geology*, Vol. XX, Nos. 6–8 (1967).

Richie, Atha Albert. "The Real Facts About Those Famous Old Hand-Drilling Contests," *Engineering and Mining Journal*, Vol. CLII (November, 1951), 84ff.

Rickard, Thomas A. "Rich Ore and Its Moral Effects," *Mining and Scientific Press*, Vol. XCVI (June 6, 1908), 774–75.

Taylor, Edgar. "Hich Explosives and Safety Fuse," *Mining and Scientific Press*, Vol. XCVIII (1909), 726.

Thompson, James V. "The Visitation Engineer Writes Again," *Metal Mining & Processing* (September, 1964), 41–43.

Young, Otis E, Jr. "The Craft of the Prospector," *Montana: the Magazine of Western History*, Vol. XX, No. 1 (Winter, 1970), 28–39.

———. "Fire in the Hole! Evolution and Revolution on the Western Mining Frontier," *American West*, Vol. VII (July, 1970), 15–19.

Unpublished Documents and Interviews

Interview with Ross Thomas, Dolores, Colorado.

Interview with Martin Duffy, Florence Mine, Goldfield, Nevada.

Interview with Leonard McCloud, Chula Vista, California.

Interview with Mrs. Martha Hickey, Mayer, Arizona.

Rykowsky, Thomas. "History of Mining in Tombstone, Arizona: 1876–1902." Unpublished paper, in possession of Otis E Young, Jr.

[Schieffelin, Edward S.]. "History of the Discovery of Tombstone, Arizona, as Told by the Discoverer, Edward Schieffelin," [together with] "Arizona: Her Resources and Future Prospects," and "Edward Schieffelin." MSS P-D2. Bancroft Library, San Francisco.

Index